The Old Marlborough Road

The Old Marlborough Road

A JOURNEY INTO WONDER

Ken Wolgemuth

ZOLAND BOOKS
Cambridge, Massachusetts

First edition published in 1991 by
Zoland Books, Inc.
384 Huron Avenue
Cambridge, Massachusetts 02138

Copyright © 1991 by Ken Wolgemuth

All rights reserved. No part of this book may be used or reproduced in any manner without written permission, except in the case of brief quotations embodied in critical articles or reviews.

Grateful acknowledgment is made for permission to reprint the following:
Quotation from All the Strange Hours by Loren Eiseley. Copyright © 1975 Loren Eiseley. Reprinted with permission of Charles Scribner's Sons, an imprint of Macmillan Publishing Company.
Quotation from The Unexpected Universe by Loren Eiseley. Copyright © 1969 Loren Eiseley. Reprinted with permission of Harcourt Brace Jovanovich, Inc.
Quotation from The Good Earth by Pearl S. Buck. Copyright © 1931 Pearl S. Buck. Reprinted with permission of HarperCollins Publishers.
Quotation from Pilgrim at Tinker Creek by Annie Dillard. Copyright © 1974 Annie Dillard. Reprinted with permission of HarperCollins Publishers.
Quotation from "The Star-Splitter" by Robert Frost, from The Poetry of Robert Frost edited by Edward Connery Lathem. Copyright © 1923, © 1969 Holt, Rinehart and Winston. Copyright © 1951 Robert Frost. Reprinted with permission of Henry Holt and Company, Inc.

Library of Congress Catalog Card Number: 91-65384

ISBN: 0-944072-16-X

Cover painting by Deborah Brown
from the private collection of Roland F. Pease, Jr.
Copyright © 1991 by Deborah Brown

FIRST EDITION
Printed in the United States of America

For Karen and Kestrel,
with gratitude and with love

Contents

Introduction   xi

1. A Personal Mythology   1
2. Season of Glass   17
3. The Real World   27
4. Hunting for a Salamander   41
5. Rain   55
6. Bound in the Web   67
7. The Creatures of Contradiction   83
8. The Hidden Wilderness   103
9. A Lesson in Dying   117
10. October Firefly   127

If with fancy unfurled

   You leave your abode,

You may go round the world

   By the Old Marlborough Road.

HENRY DAVID THOREAU

1961 Perry County, PA
Photo by Miriam Wolgemuth

Introduction

THERE is an old photograph in front of me. It is the photograph of a mountain road, a dirt track overhung by trees and disappearing, just up ahead, into the haze of a summer afternoon. In the foreground is a boy, three years old. Farther along the road, his back to the camera, stands a young man. The man, standing tall, gazes up and ahead, where sunlight is blazing through the treetops; the boy, stooping, stares at a rock.

This photograph has become symbolic of my life. I seem these days to be standing in the same relation to my peers and the world at large as to my father in that old picture. Humankind, the light of tomorrow on its collective face, is conquering disease, probing the depths of the atom, and preparing for the colonization of the stars. Meanwhile I kneel alone, in some remnant of forest, intent upon the comings and goings of beetles.

To those who have seen me only at home or at work, my life must seem the model of normality. I manage to carry out from day to day the duties of loving husband, doting father, conscientious son, and loyal employee,

THE OLD MARLBOROUGH ROAD

but there is a large part of me that has grown unfit to play any kind of productive or responsible role in the modern world.

Left to my own devices, given a spare moment or two, I tend to gravitate toward some green and growing place, be it a park or a playground or a local woodlot. There I can be found staring up into trees, poking about in puddles or worming my way through thickets. I tend to move slowly and stop often, to pick up rocks and dig around under rotting logs. More than once, parents have clutched at their children and hustled them in a wide semicircle around me as I knelt by a trail, prodding the carcass of some dead animal with a stick.

On my walks I am apt to present a raggedy appearance. I often go out unshaven, in mismatched, tattered clothes, knowing that nature will not mind and forgetting that humanity is often less generous. Emerging from the woods, bathed in sweat, with spider webs trailing from my ears, ticks creeping up my pantlegs, or frog slime smeared across my palms, I am liable to startle those who have come out into nature merely to picnic or take a bit of exercise. I like to think of myself as a naturalist of high purpose, an explorer of the close-at-hand, if you will. But I am fundamentally and forever a straggler.

I have come to accept and even cherish this role. The straggling gazelle may be struck down by the cheetah,

INTRODUCTION

but there are those of us who prefer the cheetah to the herd. Besides, if there are dangers in lagging behind one's fellows, there are also compensations.

For I have come to realize that the mountain walks I took with my father were my first steps along another, less substantial road—a road with no beginning and no end; a road not appearing on any map; a road to be discovered primarily by those who travel alone. This road is not confined to a particular place or a particular time. I have since encountered it in leafy glades filled with the rush of water, and on dark ridges under icy winter skies. I have traced it through the backyards of small towns at twilight, and glimpsed it deep in the gleaming eye of an insect. This road is marked not by milestones but by moments, evanescent instants when the earth is revealed for what it is; when the commonplace takes on a radiant significance; when, in turning my head to the distant thunder, I feel the lurching globe suddenly catching me by the heels to whirl me through the starlit depths of space and the infinite reaches of time.

Of course these moments never last. The enchantment is soon broken. The workaday world crashes through and the road vanishes in the tall grass once again. But its hold on me is strong, and there is always the possibility of finding it again. On dark days, when the undersides of clouds are shredded on the hilltops,

THE OLD MARLBOROUGH ROAD

and the waters ripple under a capricious, cool breeze, I walk with a sense of expectation, knowing that anything might happen, that at any moment I might round a corner and see, stretching off into the mist, that weedy track—that road where things appear in their proper proportions; where the hubbub of the crowd and the lights of the city fade away, and where, in the caw of a crow or the bend of a branch, one may find ample reason for remaining here, for going on.

This book is a montage of moments on that road; a patchwork quilt sewn of the vagrant thoughts occasioned by vistas along the way.

Ken Wolgemuth
February 1991
Thompsontown, Pennsylvania

The Old Marlborough Road

> What the Boy chiefly dabbled in was natural history and fairy-tales, and he just took them as they came, in a sandwichy sort of way, without making any distinctions; and really his course of reading strikes one as rather sensible.
> —Kenneth Grahame,
> *Dream Days*

1
A Personal Mythology

WHEN I was seven years old, my best friend was my next-door neighbor, Roy, a grinning, mischievous redhead who, at twelve, seemed the epitome of wisdom and sophistication. Out of Roy's vast store of knowledge and experience came some of my first lessons in the ways of the world.

The only problem I had with Roy arose from the difference in our ages. It was not so much that he was older than I, but that he never let me forget it. Rather, he took every opportunity to remind me that he was the teacher and I, the pupil; that however much I might be learning he would always be at least one step ahead of me.

THE OLD MARLBOROUGH ROAD

One summer day, Roy decided to illustrate once and for all the gap between his abilities and my own. Producing from his pocket a strip of metal he'd found, he headed for the big maple tree at the far end of my backyard, and started to climb. Two-thirds of the way up—twice as high as my parents would allow me to go—he fastened the strip to a branch, then nimbly descended and headed for his house. As he opened the screen door he turned to me with a smirk and observed that, with luck, I might in a couple of years be brave enough to challenge his record climb.

He had at last gone too far. Roy could call me short, he could call me stupid, he could call me ugly. But he could not call me a coward. Ten minutes later I had unfastened the strip from the branch and was climbing, full of grim purpose, toward the top of the tree. At last, swaying in the breeze some thirty feet above the yard, standing on branches that even to a foolhardy child seemed hopelessly spindly and weak, praying that my mother would not choose this moment to step outside, I stretched and, with the very tips of my fingers, attached the metal band to the topmost twig of the old tree.

Roy never even attempted to match my feat (never said a word about it, in fact), and for all I know that strip of metal remains where I left it. My life since that

A PERSONAL MYTHOLOGY

day has not been dull, and yet that long-ago climb stands out as one of my most truly satisfying victories.

ROY and I grew up in the most aptly named town of Mount Joy, Pennsylvania. Though destined in later years to become a booming bedroom community for the cities of Lancaster and Harrisburg, and showing even then the first new subdivisions on its fringes, the Mount Joy of the early sixties had yet to undergo the convulsive expansion, demolition, rebuilding, and refurbishing that would accompany its later, sudden popularity and the arrival of the affluent young. Rather, it was a town that had molded itself to the shapes of families who had been there for generations; a town worked in and played over by living hands until all rough edges had been rounded off, all stiffness of character kneaded out; a town as smooth and polished as a century-old banister, as wonderfully faded and pliant and comfortable as a favorite pair of old jeans.

Running through the center of town was Main Street, Route 230, at that time a principal east-west thoroughfare noisy with city-bound traffic. But to either side of this bustling artery lay quiet residential neighborhoods. The back streets were lined with oaks, maples, and sycamores whose crowns tossed shadows across golden lawns, and whose roots lifted the bricks in the sidewalks. From behind the trees peered the gables and

THE OLD MARLBOROUGH ROAD

fieldstone towers of old houses separated by hedges, or strips of flowers, or paths that led to still, sunny nooks under stairs and balconies, where cats dozed amid the buzzing of flies.

When I was young, Mount Joy seemed to lay lightly upon the land, more like something that had grown up out of the earth than something that had been built upon it. Beyond the houses lay farmers' fields, little woodlots, streams and thickets; and between the town and the country was a narrow band of commingling. In this alchemic region the realms of man and nature met to create little landscapes that, like the children who were drawn to them, were neither wholly wild nor wholly tame. With names like the Cove, the Small Cliffs, Stone Bridge, and Monkey Jungle, these were welcoming and peaceful hideaways, generous with the gifts of nature to children: rocks for climbing and vines for swinging; the feel of mud, the sound of water, the smell of captured frogs in a yellow plastic pail.

A mile from town was a rise on whose summit stood a great beacon, presumably placed there to warn off low-flying planes. I remember riding by with my family on Sunday afternoons, and craning my neck to look up at this cyclopean being with his massive red legs and single eye. At dusk this Titan came to life and surveyed his realm. The light of his gaze swept town and countryside, and fell at regular intervals upon the face and deep into the darkened rooms of our house on Main Street.

My room was on the second floor, in a back corner, somewhat removed from those of my parents and sister. It looked out over the roof of the back porch and into the yard, which stretched from Main Street to the alley behind the house. On the far side of the alley lay the railroad cut, whose banks were thick with sumac, brambles, and poison ivy, except under the bridges where the sunlight never fell.

The yard was the center of my universe. I explored the town, of course, and my father took me farther afield—to the mountains, where he looked for deer while I dug for the remains of dinosaurs; to nearby lakes and streams where we caught huge, thrashing carp—but it was in the yard that I launched my most fruitful expeditions. There I found shades of all I had seen on more distant travels. The yard was a microcosm of the wide world, and the heart of the enchanted geography of my youth.

Innumerable landscapes were contained in that swath of grass between Main Street and the alley. It was forest and prairie, veld and moor, desert and sea, whatever the enterprise of the moment required. Next to the alley stood our trees—the maple, which we kids could climb, and a cherry, which, try as we might, we could not— and off to one side were two or three lilac bushes. The bushes had grown together in such a way as to shade the ground below, leaving it bare of grass. The space between their trunks formed a path just wide enough for the passage of children in single file. In the subdued forest-light along the path I could find ants and centipedes, and sometimes the broken blue eggs of the robins that nested in the branches.

Seated there alone, on the earth and the dead leaves, poking my fingers into the soil, looking under rocks, picking ladybirds from the stems, I felt I was on the verge of discovering something important. I felt at home. The filtered, green light that came through the leaves seemed to be the sort of light in which I was meant to live. I had read about exotic animals and faraway places, but in my yard, in the humble confines of the lilac bushes, I found astonishment in the ordinary. I followed the movements of robins and rabbits, but was drawn most strongly toward nature's lesser manifestations—toward the small creatures, the cold-blooded creatures. I reached out and touched life in all its squirming glory. In the jabbing thrust of a cricket's knee, in the velvet ripple of a caterpillar's tread, I felt a life force as strong as my own.

On summer evenings, especially, the backyard would beckon. In the twilight, plants seemed to relax, to expand, to stretch their leaves to the cool breeze and release a scent compounded of summer heat and earthy dampness, of dust and the powder of butterfly wings. As the sun set and shadows lengthened, the world seemed to lose a little of its hard reality: houses grew less angular, somehow less substantial, faintly sparkling around the edges—and magic moved in the grass.

On those evenings, curtains fluttered at windows, and people rose from their dinners to drift through dusky houses and out onto back porches. Lawn chairs creaked and screen doors opened and slammed up and down the street. You smelled the earth, and the cut grass, and the wind, which, bubbling around the corner of the house, brushed the scent of flowers against your cheek and whispered off.

The adults sat on the porch and talked among themselves while we children ran back and forth over the shadowed lawn, catching fireflies that rose and winked out on every side like some luminous effervescence in the liquid evening air. They were brilliant glimmers against the grass; dark, wing-blurred motes against the sky. They cried out to be caught and held, to be collected in jars and admired in darkened bedrooms through the night.

As a boy I strode like a demigod across the world. The laws of nature were, where I went, mere suggestions, and I found the power within myself to change the very form and function of the trees and the clouds.

THE OLD MARLBOROUGH ROAD

At the same time, however, I was conscious of powers outside of and far greater than my own.

Sometimes, on summer afternoons, a stillness would fall over the town, as if the world had suddenly stopped what it was doing and stood to await the coming of a stranger. Bees stopped their humming, leaves hung motionless, and you could almost hear the sizzling of the sun as it beat down upon the yard.

Suddenly the light would fade from the grass and the sky, and the maple tree would turn from green to silver. Something in the sight of that silver tree awakened a part of me left over from man's time in the forest. I stood and sniffed at the air like a wild creature. I was alert and all alive. God was passing his hand over the earth as a magician passes his hand over a hat. The sky grew dark, and I waited for the magic.

As the wind picked up, and a muted rumbling fell from the northwest sky, one or two big drops would make black spots on the sidewalk, and I would head for the house, fast. I would kneel just inside the screen door, nose to the mesh. As the storm broke over the town I would be bathed in the most evocative of scents, one concocted of water and dust and screen, and, if the timing was right, of supper simmering on the stove. Outside, rain splashed, thunder echoed, and every few seconds the lightning's bright fingers snatched up the

A PERSONAL MYTHOLOGY

yard I knew and left in its place an unearthly landscape of silver and coal.

Through the rain's gray veil I would glimpse movements—dim, shifting shapes that somehow contrived to vanish with each flash of lightning. I knew what they were, for we were old acquaintances. I had sensed their presence many times—while trick-or-treating on Halloween, or walking home along empty streets from choir practice. I had heard their passage through the fallen leaves, and caught the sound of their voices on the wind. They were the masters of the dark wet side of nature—the shy ones with no names, the shimmering, insubstantial beings who transacted their business at the edges of our lights. They were as real to me as worms and crickets, rock and iron, wonder and dread; their habits were to be taken into account in any serious study of nature.

Like the boy in Kenneth Grahame's story, I made no distinctions between what are called fantasy and reality. I believed in monsters, and I believed in magic. And why not? A dragon could be no more wondrous than a firefly; the transformation of a frog into a prince no more improbable than the shaping of a caterpillar into a moth. The natural and the supernatural, the overt and the hidden, merged with and interpenetrated one another to form the fabric of the universe in which I dwelt;

THE OLD MARLBOROUGH ROAD

science and superstition were to me two sides of the same coin.

WHEN I was young my paternal grandparents lived ten miles away, in a haunted house at the edge of the world. On one side of this dwelling was a parking lot—a vast, empty, echoing place, especially at night, under the moon. Beyond the lot was a church, and beyond that the lights of human habitation. On the other side of the house there was a bit of grass, and then nothing. I can picture nothing there—no houses, no fields, no roads—just a shimmering void, the edge of the world.

The house itself was old and dark, with a facade of blackened brick shadowed by two massive and gnarled trees. Friendly, watchful beings, they guarded this, the last human outpost on the frontier of the land of shadows.

Inside the house were corridors and stairs, and shaded rooms where curtains billowed in the breeze, admitting dusty rays of light. From the outside the house seemed too small for all the chambers and passages that unfolded within. There was a room upstairs, unused and bare of furniture, into which I never stepped, certain it was an illusion that would dissolve beneath my feet.

One summer, my cousin and I spent a weekend with

A PERSONAL MYTHOLOGY

our grandparents. We kids were alone in the house with Grandma, Grandpa having gone out for some reason and not yet returned. We had just finished our supper, and the sun was going down behind the trees, when a great, green-black insect appeared on the front stoop.

The beast's wings quivered as it crept mechanically toward the three of us huddled in the doorway. We all looked on, not quite knowing what to do, having never seen a bug so big or so menacing. Vaguely recalling the picture in one of my books on nature, I formed and spoke the word *cicada*. But the fact that the creature had a name, that it was indeed known to science and not something newly formed from the night mists, did not render it harmless in the eyes of a woman whose curiosity about nature took a back seat to her concern for the welfare of her grandchildren. Grandma reached for an aerosol can.

As the sun set and shadows crept up the face of the house, the only sound was the hissing of the spray can; the only movement, the fitful crawling of the bug. Beneath the deadly blast of spray, it slowed and it paused, but then it began to crawl again—straight for the door, straight for me. It was not until the can was empty, until the spray had fizzled and then ceased entirely, that Grandma lifted her finger from the valve. And only then, as the last of the light slipped from the roof, did the bug stop, teeter, and finally fall onto its side. Its legs twitched for many minutes as the rising night breeze swept away the stench of the poison.

The bug did look dangerous; it did look sinister. Yet this killing disturbed me. Such things might be done with impunity elsewhere, but there, on the brink of reality, the most mundane act could have unforeseen consequences. Beneath the brow of that creaky house, in the blue shadow of a summer evening, the insect had a look of purpose about it. It seemed driven to survive. More than that, I had the feeling it had come with a message—a message for me.

This feeling was nothing new. The world was full of messages in those days. Perhaps it was quieter then, and easier to hear. Perhaps *I* was quieter then, and knew better how to listen. At home, at night, I would lie in bed, listening to the crickets creak in the yard. I shook to the passage of passing freights while the beacon's eye, revolving on the hill, alternately lit the world and left it in darkness. I heard a voice calling through nature. The words were unclear, but the tone was one of authority, and I ached to understand. From my window I looked out over the darkened planet and I sensed the magnitude of the mystery of which I was a part.

I NO sooner sensed it, it seems, than I lost it. I was distracted. I turned around once, for just a second or

A PERSONAL MYTHOLOGY

two, when I was about twelve; when I turned back, ten years had elapsed and everything had changed.

Not only were cliffs lower, streams narrower, and houses smaller, but the color had faded from the sky and the grass, and the brilliance was gone from the sunlight. The house at the edge of the world had been demolished, and the place where it had stood was discovered to lie, inexplicably, well within the bounds of a small city. As I aged, my gaze had grown wider and the world had shrunk beneath it. In some ways I saw more and I saw farther, yet what I saw was flat and lifeless and ordinary.

The senses were still there. The eyes, the ears, the fingers were all in working order. But the wiring had changed; circuits had shifted. Almost all channels now ran to the brain, the cool and analytical brain. Pathways to the heart, to the soul, had begun to corrode, and communications, with rare exceptions, were closed. More often than not, I failed to see the wonder that was an anthill because it arose amidst shards of glass on a vacant city lot; failed to hear the piping of Pan over the clamor of men and women.

But I remembered then, as I remember now, the life I had lived before. What's more (and this is my problem) I knew it to be real. If I could just let it all go . . . If I, like so many others, could just write off

THE OLD MARLBOROUGH ROAD

my boyhood experiences as so much childish nonsense—life would be much simpler now. But I cannot. I simply cannot bring myself to concede that what I saw then, what I felt then, was illusory, something meant to be outgrown.

On the contrary, childhood shimmers on the horizon of memory like a lost continent where great deeds were done, great battles fought, great lessons learned; where signs were read in the shaping of the clouds and the voices of the gods were heard in the rain. It is the lens through which I have viewed, the standard by which I have measured, all subsequent events. It is my *Odyssey,* my *Iliad,* my *Aeneid,* and my *Paradise Lost.* It is a personal mythology. Never since has the world seemed so big and the light so bright; never since has each day dawned so full of promise, and each night fallen so cloaked in mystery.

I've long held an image in my mind, a memory so powerful that it hangs on through the years as both an inspiration and an admonition: It is the first truly springlike day of spring, sunny and warm, with the scent of the earth strong in the air. The Boy stands with his mother in a back room of the house on Main Street.

"Can I go outside?" he asks.

"Oh, I suppose."

"Do I need my jacket?"

A PERSONAL MYTHOLOGY

"I guess not."

And so a jacket is flung aside, and a screen door is thrown open. And there the picture freezes: the Boy, laughing, his arms in the air, is framed in the doorway, in a rectangle of blinding light. Beyond, within the light, is a yard, and a town, and a world. It is a world of limitless space and unqualified freedom, where the commonest of events has a private meaning and where nature, seen truly, is at once the stuff of science and of dreams.

That world is a difficult one to find, lately, yet the power of this image assures me that there is a way back in.

So I am retracing my steps, both in space and in time, following the crumbs I dropped so long ago, a trail that leads not out of but back into the dark wood, where birds help to guide me by day and fireflies light my path after sundown. I sense that if I follow them long enough they can take me where I long to go. I trust that one day, it may be tomorrow, I will find myself on a windswept hill; that I will look down into a distant valley; that I will glimpse, in the top of a tall tree, the gleam of sunlight on old metal, and know that I have returned at last. And there, somewhere, on sidewalks of brick or in twilit, lilac-scented yards the Boy, forever seven, still runs and plays—still laughs and cries and wonders and dreams in a hidden dimension where the rain has always just now ceased, where the wind smells of cut grass, where jackets are not needed, and where Saturday stretches out ahead of him forever.

I can hear the night frost split stones in deserts. Men are softer than stone, much softer.
—Loren Eiseley,
All the Strange Hours

2
Season of Glass

LATE November. I stand on the shore of a lake after sundown. There is a hint of violet in the sky and in the water, but the woods at my back are dark. The air has turned cold and heavy, and I can feel winter, like a great silence, lying in wait just over the hills to the north. In front of me, on the shore, weeds rustle softly. Their leaves, dried and curled, trace lazy semicircles on the water. In the shallows, just a few yards out, a muskrat house stands proudly. A living being has here fashioned itself a home, has taken the dried grasses and the twigs and the mud, and made this sturdy little mound, a place of warmth and coziness in a world turning cold. I shiver as the light fades from the water and a star is kindled

THE OLD MARLBOROUGH ROAD

over the distant ridge. There is, in the muskrat's lodge, a certain fitness, a harmony that fills me with longing. I envy this dwelling, and its maker, so very much at home in the world.

IN my transformation from child to adult, nothing suffered more than my relationship with winter. We used to get along so well. There was a time, I remember, when I looked forward to it—to the snow-days, to the forts and tunnels, to the pitched battles among the drifts. I still remember a December night—I couldn't have been more than five—and how I stood on the sidewalk and watched, mesmerized, as snowflakes swirled out of the darkness overhead, and turned red and green as they swept past the electric Christmas candles on the telephone poles along Main Street.

I remember, too, how as a teenager I walked alone one winter's day in a state park; how, as I entered a deserted campground, the end of a gnarled limb resolved itself into an owl and launched itself up to soar silently around me, drawing ever larger circles on the sky until it vanished at last behind a ridge. I remember striking out across the frozen lake, not knowing, and not bothering to question, whether or not the ice was safe. I remember the exhilaration of crossing that glassy, windswept plain alone, and of safely reaching the other side, and of daring to return the way I had come.

Most vividly of all I remember, that same winter, spending a weekend with a friend at his grandfather's cabin in the mountains of northern Pennsylvania. We took a walk one night, down the winding, snow-covered lane and from there into the forest. The snow crunched underfoot, and stars by the thousands glittered overhead. Without flashlights, we inched our way around the base of a cliff and descended to a brook, still gurgling in spots though holes in the ice. We lay on our backs on a great flat rock. As the heat drained from our limbs, we listened to the water, and stared up through the hemlocks at the stars.

We climbed a snowy slope through the trees and emerged into a clearing on the side of the mountain. A weathered old shack, its timbers falling away, stood there deserted, gleaming faintly against the evergreens. Stepping onto its rickety porch we turned and looked out into an ice-blue valley. The stars dangled just out of reach, danced at arm's length on the dense, cold air. Crystals of ice and snow, like strewn gems, caught the light, and shattered it, and flung the pieces in all directions. It was as if the earth had put on a blue sequined gown, or we had stumbled into a sapphire sea shot through with diamond fishes. The light from that hilltop flashes in my head to this day.

BUT winter is not what it used to be. That I now have the obligations of a responsible adult—that to go out is no longer a choice but a necessity—is part of the reason, I'm sure. But there is something else, too, about the winter, something that puts an icy knot into the pit of my stomach whenever the forecast calls for snow and sleet. There is something in the whisper of the falling flakes that speaks not merely of inconvenience but of trouble ahead; something in the bitter cold that turns purpose into paralysis and conviction into doubt.

In the summer I am pliant; in the winter I am brittle. In the summer I am flame; in the winter I am glass. I fear a stronger wind, a colder night, a sudden shattering.

If I walk at all in the winter, these days, it is only for short periods, more because I feel I should than because I want to. I do not often think, anymore, of venturing out onto frozen lakes.

ON a February afternoon, I set out for the woods. The wind blows relentlessly and capriciously, veering, it seems, to keep always in my face. The woods, spare and ragged, offer little protection from its blasts. Bare branches rattle overhead; old trunks sway and creak. In a clearing, dead grasses bob crazily, pitch first to one side and then another, flatten and rise according to the whims of the gusts. The withered leaves and blackened berries clinging to a honeysuckle vine whir and chatter. The sun is shining, but its light is diffuse and impotent. The woods have no scent, unless *cold* is a scent. I walk quickly and seldom stop, in an attempt to stay warm. I tramp up and down hills, chin to my chest, absorbed in thoughts of home and hot coffee.

Little is stirring. A few hardy squirrels rattle the dead leaves. Small birds flit from tree to tree, stopping now and then to peck at crevices in the bark. A dozen or so overwintering geese pass in a ragged skein high overhead.

The trail is frozen, and the soil gives with a slight crunch at every step. The lake is frozen, too, but the ice is very thin. It is ribbed and shot through with glassy spicules, and breaks readily when poked with the end of a stick. A small stone, gently lobbed from shore, cuts through with a sound halfway between a whistle and a swoosh. Indeed, the ice is so thin, so ribbed, so elastically attached, that it has become a musical instrument, a fragile drumskin stretched between the two shores. A stone or broken branch skipped across its surface fairly sings. The sound is ethereal, airy, the sound of a wet finger on the rim of a crystal wineglass. It rises briefly on the air, and, humming, is swept off through the deserted woods.

I've walked this path a hundred times, but today everything is unfamiliar. Every contour of the woods is laid bare. Slopes and boulders that were invisible in the summer take me by surprise, appearing unexpectedly, as it were, just a hundred feet away. Old stone walls that

THE OLD MARLBOROUGH ROAD

I had never before noticed are conspicuous, angling up the leaf-strewn slopes from the lakeshore. The world is suddenly a hard and flinty place, all angles, edges and points—naked and strange. Winter is skeleton time. It is life stripped down to the minimum, with nature's bones exposed.

As I turn a corner on the trail, a great gust of wind plows through the forest like an invisible train, whistling in the branches and scattering the dead leaves. It seems to pass right through the fibers of my clothes; it rasps my face like a cold razor. I try to make a note in my notebook, but my hands are not working, and the pen falls to the ground. I can barely see through wind-drawn tears. I am starting to shiver violently. Now deep in the woods, I turn suddenly and hurry toward my car.

Swaddled in layers of wool, down, and nylon, I stumble along the rutted trail on benumbed feet, with dripping nose and watery eyes, hands dangling uselessly at my sides. Yet tiny chickadees flock about me as I hurry on. With grace and confidence, they flit and they circle. They perch on nearby twigs and whisper among themselves, tsk-tsking my foolishness in venturing out, a beast so unprepared for the cold. They, it is obvious, are at home, and welcome. They, it seems, are somehow invisibly connected to a deep-hid wellspring of warmth from which I am excluded.

Reaching my car at last, I fumble for my keys, then, hand shaking, struggle to work the lock. Once inside, I slam the door. I find reassurance in its solid "thunk." Later, safely home, showered, and with a hot meal inside me, I curl up on the couch, in the glare of the television, and contemplate with smug satisfaction the purr of the furnace in the basement. We are intelligent, inventive, and adaptable, I tell myself. That is how *we* survive. Our technology is our refuge from the cold.

But even as I sprawl in comfort I am aware of the wind, still howling outside, and aware, too, of the feebleness of my argument. These homes of ours, these massive structures of timber and brick and steel and stone—if they are a testament to our intellect and our adaptability, then they are a symbol, too, of our frailty. These proud boxes upon which we so depend are themselves dependent upon their pipes and their wires and their fuel: pipes that can freeze, wires that can fall, fuel that can run out. Nowhere do our foundations rest on truly *solid* ground. I may put brick upon brick and board against board, build higher, thicker, stronger, yet my sleep on these winter nights will never be as sound and untroubled as that of the muskrat in his mound of sticks. For his home is but a resting place fashioned in league with nature, and mine, clearly, is a bulwark thrown up against her.

We find in the winter what a shaken Thoreau found

THE OLD MARLBOROUGH ROAD

amid the trailing vapors and grim stones on the summit of Mount Ktaadn: we find a nature "not bound to be kind to man," a nature whose solicitude we are denied, or perhaps, in our arrogance, have forfeited. The world in winter is a cocoon, or a cyst. It has shrunken in toward its center, has wrapped tightly around itself and folded its tender, green, and growing parts within. It shows, without, a hard, impervious coat, and we, it seems, have been left on this silent surface alone.

The lesson of winter is that the rest of nature is somehow looked after, while we are fated to look after ourselves. The terror of winter lies in the thought that we may not be equal to the task.

IN a matter of weeks, I know, there will come a change. Perhaps there will still be a few patches of snow outside, still a few spots of partially frozen ground. But I will step out back and the air will be warm, and will smell of earth. From somewhere will come the sound of water trickling. A group of mourning doves will play a recorder symphony in the trees; others will fly past on whistling wings. A cardinal will sing a partial and tentative song. When I take a step, the ground will part with a wet, kissing sound, and come away with my shoe. I will hear the buzzing of a fly, and I will smile. I will stand straight, and look out over the world—and it will be my world again. I will breathe deeply, thrust out my chest, and strut with purpose across the slowly greening lawn.

IN the meantime, I sit on the couch and wait. The wind buffets the walls. On the table beside me, the lamp begins to blink. A bare branch is tap, tap, tapping on the windowpane.

> Blessed is he who has found his work.
> —Thomas Carlyle,
> *Past and Present*

3

The Real World

I WAS about fourteen, I think, when I first began hearing from parents and teachers about a fearsome place known as the Real World. I would be daydreaming out loud, rhapsodizing about how I was going to do this and that, how my life was going to be different from theirs, and they would just nod, and chuckle knowingly, and say, "Wait till you find yourself in the Real World." I never worried very much about this prospect, though, being fairly certain I would be able to recognize the outskirts of the cold, gray landscape they described, and thus avoid it. And I did just that, too, for longer than most.

Immediately after leaving college, I joined the Peace

THE OLD MARLBOROUGH ROAD

Corps and spent the next two years in a hammock on the back porch of a mud house in tropical West Africa. When I left the Corps, I came home and took a job as a ranger at Gifford Pinchot State Park in Pennsylvania. For three seasons, from April through November, I watched the world through the windshield of a patrol car, occasionally stopping to take a short walk or to tell some miscreant to leash his dog.

These years were a time of moving slowly and looking around, of growth and spiritual enrichment. That is, they were nothing like what most people know as real life.

But things gradually began to change. The ranger job, as I have said, was seasonal. By the end of my second year I had finally left home for good and moved into a place of my own. Paying rent and buying food meant that I was actually going to have to work during my winter layoff. With no better prospects, I signed on as a laborer for a temporary-help agency.

I spent two weeks shampooing the upholstery of cars, a few days directing traffic for a road-repair crew, one morning unloading a truckload of dented cans and leaky jars at a discount food store, and a week in a local factory, punching holes in aluminum tubes destined to become the frames of baby strollers.

Except for the traffic detail, it was hard work, and not terribly exciting, but for a while I was able to look at it in a romantic sort of light as one more step in my education, a broadening of my horizons—an experiment, if you will. Then one day in late winter came a call that at first I did not take seriously. I asked the woman to please repeat what it was I would be doing, and I was still shaking my head in disbelief as I drove to the job site. To pick up litter for two days was a job I could understand. To pick up litter for two days, in the cold of early March, was a job that sounded uncomfortable. To pick up litter for two days, in early March, at a *landfill*, was a job that was just too unspeakably absurd to be real.

Situated on an unprotected rise on the outskirts of the city, the landfill would have been a depressing spot on the brightest, balmiest, and most flower-scented of May afternoons. At dawn on an overcast and bitter cold winter's day, it was so stark and unforgiving a landscape that I would not have been overly surprised to find, scrawled on some rock, "Dante was here."

A foreman led me and two others to a vantage point overlooking the entire operation. We looked out on an expanse of raw earth and dead weeds that extended from where we stood to the top of a ridge a quarter of a mile away. A bitter and howling wind swept across a distant mound of cinders, turned black and gritty and went hissing over the barren slopes. Loose papers and tatters of plastic tumbled past, lodging against the fence

THE OLD MARLBOROUGH ROAD

or sailing over it onto adjacent roads and fields. In the distance, bulldozers and trucks grunted like huge, rooting animals; above, flocks of gulls spiraled and shrieked. Swinging his arm in an arc to take in the fence, the roads, and the fields, the foreman outlined our simple task: "If it's blowing around, bag it." As I set out across the muddy slope it occurred to me that this was not exactly how I had envisioned living my life.

I moved along the fence line in the dim light of early morning, chasing down runaway papers, disentangling old cotton balls and bits of tissue from the weeds, stuffing everything into the sack at my side. The flying cinders scoured my frozen face, lodged in my hair and in the folds of my clothes. My nose began to run, and when I wiped it my handkerchief came away streaked in black. By the middle of the morning I was sore, filthy, and numb with cold. By the end of the day I had had a long overdue revelation: life, I saw, would not at every moment be a thing fraught with meaning and transcendence; often it would be dirty, and tedious, and just not a lot of fun. I had grown complacent and careless, and had stumbled smack into the middle of the Real World. Funny, I thought as I looked out over the landfill, it was just as I had pictured it.

HOME at last, I scrubbed off the worst of the day's grime, put a pot of coffee on the stove, lit a cigarette,

THE REAL WORLD

and eased myself down into a chair in the little trailer I called home. The only light was the soft blue glow from the gas stove; the only sound the cozy, liquid babble of the percolator. I couldn't get the landfill out of my mind, but in the darkness, as I relaxed, a change came over the landscape I saw. The mud and the weeds gave way to a grassy plain dotted with tracts of forest. From within the forests came not the chugging of bulldozers and trucks but the snuffling of great, scaly beasts. The wind blew, but the hiss of cinders was replaced by the snapping of the pennants that flew from a distant stone turret. Ah, yes, this was more like it.

THE trouble with the Real World, I decided, was that it had become entirely too real. It used to be that one could pass through the castle gates and enter the dark wood in search of adventure. There were maidens to be rescued, villains to be thwarted, quests to undertake. But not anymore. The world had been explored and mapped to within an inch of its life. The white spaces had been filled in. Lands of magic and monsters gave way to quite ordinary mountains, plains, and deserts.

While our planet shrank, our universe grew. What had been a cozy system of nested spheres (with us at the center) dilated into a fearsome void, measured in light-years, where galaxies moved like huge amoebas and the ghosts of dead stars devoured matter and light. The

THE OLD MARLBOROUGH ROAD

cosmos was shown to be a cold, dark place growing ever more thin and lonely as everything we could see flew away from us and from everything else at unimaginable speeds toward unknown destinations.

As we explored and charted our planet and its environs, we flushed the gods from individual rocks and streams and drove them to the groves, to the temples, to the summit of Olympus, and finally off the planet and into the depths of space. It's no wonder they seldom speak to us anymore. In today's world, burning bushes merely turn to ashes, and whirlwinds do not stop to offer advice and guidance as they tear through trailer parks. The gods left us their books, of course, but there are many of them, from many times and places, and they have an unfortunate tendency to contradict one another, leaving us confused. What's more, too often they tend to direct our attention to the next world, when what we need is some sound advice on what to do here and now.

With every advance in our knowledge there came a corresponding decrease in our sense of destiny, of having a role to play that was equal to the universe we inhabited. The dragons had all been slain. Or, if they had not, then they no longer frequented our neighborhoods.

Existence itself, once nothing short of a miracle, was shown to be the result of blind and directionless chemical and physical processes. It is like this, says the scientist, stepping to the blackboard: "Take an eternally existing universe and set it to expanding; allow a couple of generations of stars to coalesce, to burn and to explode; permit water to condense on the surface of a suitable planet; let fly a few properly aimed bolts of lightning; wait some billions of years; and before you can say 'I think, therefore I am,' you have thermonuclear explosives, microwave pizza, civilization—utterly explicable; damn near inevitable." According to Darwin, the highest, most fundamental purpose to which each of us can aspire is to produce more babies than the next person. And while it sounds attractive in theory, the pursuit of competitive reproduction simply does not silence the tiny voices that tell us we are cut out for bigger things.

What we want is a life worthy of our dreams and commensurate with our perceived but unrealized potential. We need to be reassured that we are a part of the world and that we have a part to *play*, an active, conscious, personal destiny to fulfill; a switch to throw; a purpose that goes beyond mere survival and spawning. We want to be shown that we are threads without which the Great Tapestry would be incomplete; cogs whose loss would cause the Grand Machine to grind to a halt. Alone, at night, in the silence of our rooms, we stare at the ceiling and dream of greatness. The work-

THE OLD MARLBOROUGH ROAD

aday world is full of saints, seers, prophets, and crusaders waiting for the circumstances that will allow them to reveal their true natures, would-be heroes needing only the last link to be forged in the chain of events that will make the world ready for their appearance. But life as we know it is rarely magical or heroic. Often, it is uninspiring, predictable, safe, and dull. And so we capitulate, trade in our lives for careers, our dreams for ambitions. We climb the corporate ladder when we should be scaling the battlements, and compete for the annual sales award when we should be seeking the Grail.

I REMEMBERED, as I poured my third cup of coffee, that the garbage had to be taken out. This is what we who live in the Real World do, I thought. Not bothering to take a jacket, I went out and dragged the bulky plastic can down the driveway to the curb. Water was running in the little drainage ditch at the side of the road. It made a soft, pleasant gurgle that I somehow found reassuring. The air, though bitterly cold, felt good on my arms and face. The sky had cleared, and the stars seemed unusually close and brilliant. I found myself entranced. I kept staring up at those glimmering lights, up through the bare branches of the trees, as the little stream gurgled nearby. They taunted me, somehow, by their mere inscrutable presence.

THE REAL WORLD

Mentally, I ran down some of the things I'd learned about the heavens: that there are perhaps one hundred billion galaxies out there, each containing on the order of one hundred billion suns; that all of these galaxies are flying away from us and from each other as the very fabric of space expands; that recently what seem to be planetary systems have been discovered surrounding nearby stars; that the existence of such systems makes it possible, perhaps likely, that as we gaze up into the night sky others, elsewhere, are gazing back. I thought of quasars, pulsars, neutron stars, red giants, white dwarfs, and black holes; of space-time, and gravitons, and the odd things that happen when one approaches the speed of light. It struck me that the average interested citizen of today knows more about the universe in which he or she dwells than did most specialists of a hundred, or even fifty, years ago.

Without question, our lives are richer for these facts. And yet, I thought, it is not the facts that take my breath away when I look up at the stars. It is not the facts that leave me open-mouthed and staring, reluctant to go inside even though I am on the verge of freezing. No, it is something else altogether.

I ARRIVED at the landfill the next morning and found things pretty much as I had left them—actually a

THE OLD MARLBOROUGH ROAD

little worse, for the wind had summoned up and scattered a fresh crop of trash during the night. As soon as I had worked my way out of the foreman's line of sight, I sat down to have a good look around me. Here were water cycles and food chains, birth and death, decay and regeneration, the mixing of air masses, the interchanging of day and night, and the falling of the earth through space around the sun—all the mechanics that science has demonstrated to us. Here, too, in the chugging of the great machines, in the waste and the rot, in the mounds of filth and rusting metal, were echoes of the Preacher sighing, "Vanity of vanities; all is vanity." Yet there was also something more. In the play of the light and the howl of the wind and the screaming of the gulls there lay a beauty and a majesty that transcended the dogma of the theologian and the mathematical scribblings of the scientist to touch a part of me as ancient and as innocent as the first thinking being who looked up through the trees at night and asked, "Why?"

I remembered an incident that Loren Eiseley related in *The Unexpected Universe*. One night, Eiseley and a poet friend attended an open-air opera. As the two watched the drama on stage, a great moth appeared and began to wheel about the arc lights above. The poet grabbed Eiseley's arm and whispered:

"He doesn't know. . . . He's passing through an alien universe brightly lit but invisible to him. He's in another play; he doesn't see us. He doesn't know."

If our world seems ordinary and dull, and our lives seem without discoverable purpose, perhaps it is because we, too, circle our lights. Moving from the one called "science," to the one labeled "religion," to the one known as "philosophy," we are blinded by the brilliance of our own accomplishments. We forget that much is still unknown; that much may indeed be unknowable. For the universe we encounter is filtered through human capabilities that may allow us but a partial view of reality. Science, for instance, depends on the information we get from our senses, which may or may not be complete. Even divine revelation must be couched in phrases we can understand, and may lose something in the translation. The Bible, the Vedas, the Koran, the Tao Te Ching might, for all we know, be the cosmic equivalents of children's books.

We are looking in the wrong direction. What we seek will be found not in the light but in the darkness, not in science or conventional religion but rather in that which predates and has given rise to them both.

We exist somewhere between unconsciousness and omniscience, and either extreme would be a monotonous and joyless state. It is the darkness as much as the light that imparts to life its greatest sweetness and depth; it is our ignorance as much as our knowledge that bestows upon us our greatest gift—a unique and boundless capacity for wonder. And it is in wonder, I think, that the purpose we seek may lie.

Every artist, it is said, must work for himself, to express that which is deepest within him. Yet every artist is also a communicator, requiring or at least desiring a public to ponder, to reflect upon, to admire, his creation. Maybe the architect of the universe is no exception. Maybe that's why we're here. Maybe we're the public.

It might not be the grand purpose for which we had hoped, the starring role we had envisioned for ourselves; it might be hopelessly simplistic. Still, I've not heard a better suggestion, or one that meshes so well with my natural inclinations and abilities. For me, at least, it will suffice. The day-to-day exercise of wonder is a means by which I in all my smallness am connected to the whole of the cosmos. The exploration and appreciation of the world is something I can *do*. It is a job that transcends all conventional employment; a profession that can be practiced from behind a desk, from a sickbed, from a penthouse, or from a tenement. And if the pay is not the greatest, the fringe benefits are incalculable.

THE REAL WORLD

Working faithfully at this task has lent a certain dignity, even a certain magic, to my life in this so-called Real World. It has kept me interested, and hopeful, and sane. More importantly, it has kept me humble before a universe whose dimensions we may have charted but whose potential we may often underestimate. It keeps forever fresh in my mind the image of a moth and a light and a great, pulsating darkness. It keeps me ever mindful, when I begin to strut and swagger through my little world, that somewhere, out there, in some vast, starlit amphitheater, an unseen audience may be laughing.

> To see clearly is poetry, prophesy, and religion—all in one.
> —John Ruskin, *Modern Painters*

4

Hunting for a Salamander

SPRING has worked its transformation on the woods. A few weeks ago the oaks and poplars laid a stark gridwork across the sky, and the forest floor was brown and bare, plastered with soggy dead leaves. Today I look up into soft billows of new foliage, and the ground is obscured, sunken from view under rippling leafy shallows. Mayapple, jewelweed, poison ivy, ferns: recently small and inconspicuous, they have unfurled, stretched, and merged in the dampness under the trees.

This is the time for wandering at will; for steering by impulse; for leaving the footpaths and following the deer paths. In the summer and fall, vegetation is too lush and tangled, or too seedy and spiny; in the winter, the slap

THE OLD MARLBOROUGH ROAD

of a sprung twig against a frozen cheek is too much to endure. To abandon the trail is to discover new marvels in old haunts. Not grand marvels, but small ones: columbines nodding over mossy shelves; massive old stumps, majestic in decay; leaf-filled puddles, opaque and mysterious, hinting at the concoction of new life. To stay on the path is to restrict yourself to a small portion of the land. To strike out into the weeds is to know how large the world really is, to know something of exploration and discovery, if only on the humblest scale.

Today is a good day. I simply feel good. Anxieties slough off into the wind, sift through pine needles and grassblades, and drift away. Freed from my parka I feel weightless. My arms and legs swing almost of their own volition. I believe that I've finally thawed, after what seemed an interminable winter.

Most of my walks are simple rambles, unscripted and without goal, but this one is different. For once I move with a purpose. There is a beast in these woods. He is a wily and secretive beast, but he must be here, and I mean to find him. I walk beside a shallow stream, taking note of quiet pools among the rocks. Kneeling, I dip a tentative finger into the water, testing its temperature, then idly turn over a couple of small stones on the bank, and poke about in the mud beneath them. A young woman has stopped a few yards away on the path. She eyes me, leery and yet intrigued, as I stand, brush bits of mud from my pants, and make an entry in my notebook. At last curiosity overcomes suspicion. "Excuse me," she begins, "but may I ask what you're doing?" A fair question, so I answer: "I'm hunting for a salamander." There is a pause, and then, "Why?"

Why? I don't know—I'd never really thought about it. It seemed appropriate, somehow. I mean, if you call yourself a naturalist, then this is the sort of thing you do.

The woman walks on, but her question nags. Why indeed? I find myself thinking. Why hunt for a salamander? Why, for that matter, walk at all? Why the week-to-week combing of the same trails through the same woods? Why the lists of the birds and mammals seen, the dates of the blooming of flowers? Why the cryptic jottings about the temperature, the light, and the wind? Why the descriptions of unknown mushrooms or insects, species to be looked up and identified later? Why, in short, the need to know what is going on here when the one certainty is that whatever is happening will happen whether or not I know about it, or care?

SUSPICIOUSLY, I glance at the mud-stained notebook I hold in my hand. There are more like it at home, only larger, the final resting place of all the observations made in the field. They sit there on the shelf like so many rocks, their pages the strata in which

THE OLD MARLBOROUGH ROAD

are embedded the particulars of walks past. They are something of a mystery even to me. I rarely refer to them. Yet I would not dream of throwing them away. On the contrary, I have come to look upon them with a kind of reverence.

I have made it my business to collect natural facts—to act as chronicler of the woods—yet it is not the facts, as such, that I am after. That would be a fool's errand. The United States is well-trod ground. The animals and the plants, the geology, the weather, and the other phenomena I encounter on an hour's stroll through the local woods have been observed and written about since the time of the Plymouth colony. If my purpose were merely to collect information, I could do it, and do it more efficiently, in the library.

Nothing I see has not been seen before. Nothing I jot down in my notebooks will advance human knowledge in the least. Many naturalists, I am sure, are fired with high scientific purpose. They want to chart the progress of seasons, or put their names to new varieties of slime mold, or bring clarity to the reproductive behavior of an obscure species of moth—all admirable goals. As for me, though trained as a scientist I admit to having little purely scientific curiosity. Moreover, I haven't the patience to undertake the kind of sustained, directed, and specialized observation that carries weight in professional circles. Hymenopterist Howard E. Evans says

HUNTING FOR A SALAMANDER

in his book *Wasp Farm* that there ought to be a specialist in everything. I like to think of myself as a specialist in generalities.

I MOVE on through the woods. The day is comfortably warm; the skies are clear. Beneath the trees, with their delicate and translucent new leaves, the light is unusually bright; every rock and twig is sharply defined and gleaming. I kneel and grasp the overhanging edge of a flat stone, feel its flinty teeth cold against my skin. I lift it up and smile at the sudden commotion I have caused among the ants and the sowbugs gathered beneath. I look closely, follow individual insects, take note of their starts and stops, the movement and effortless coordination of their legs. Almost on my belly, I am staring at the creatures like a man whose eyes only yesterday knew light for the first time. And that is not so very far from the truth.

WRITERS as diverse as Thoreau, Joseph Wood Krutch, and Annie Dillard have observed, rightly, that seeing is a much more difficult task than is popularly believed; that to see something is more than just to notice it; that seeing is not so much an inborn ability as it is a skill that must be learned, and practiced. They developed techniques of seeing, little tricks designed to wake up their senses to the presence of the world.

THE OLD MARLBOROUGH ROAD

Krutch, for instance, would speak to the animals and plants he encountered on his walks, call them by name in order to impress on his consciousness the fact of their existence. Dillard tells of maintaining in her mind "a running description of the present"—a strategy similar to Krutch's but perhaps more easily put to use by those of us who frequent crowded parks, where bystanders might be alarmed by the sight of someone passing the time with a mushroom or a snail.

I hit upon my own technique entirely by accident. I was walking in the woods one day, seeing little. The deer were hiding, the birds had flown; even the chipmunks declined to come out of their holes. Pausing next to a spicebush, I noticed that one of its leaves had been folded neatly in half, exposing the pale underside. Idly, I flipped it open. There, on a pad of silk, rested a creature I had known only from books: the green-and-yellow, humpbacked, and eye-spotted caterpillar of a spicebush swallowtail. Instantly, the woods that had seemed so drab and lifeless were filled with the promise of hidden treasures. I began rolling logs and flipping rocks. I found a spider and let it crawl across my hand. I knelt and regarded things from a rabbit's point of view, and found whole worlds at work among the weeds. When I stood, I discovered that the birds had come back, and it struck me that perhaps they had never gone.

HUNTING FOR A SALAMANDER

I had stumbled upon my own Rule of Vision: in order to see, it is first necessary to touch. It is necessary to grasp and to poke and to pry; to get grass stains on your knees and dirt under your nails; to bow down and put your hands on the *details* of the world. When the woods seem barren and empty of wonders, the fault is not in nature but in us. There is always more to see. I am convinced that a human life is insufficient to exhaust the possibilities of even the smallest woodlot.

It is only when we have learned to see that we realize all we have overlooked before. The act of perception pushes back the boundaries of existence, and changes the laws by which we formerly had lived. When I was younger, for instance, I used to marvel at the sudden coming of spring. One day, after months of cold and snow, I would wake to find the lawn green and speckled with dandelions. The season had not progressed, it had detonated. When at last I learned to see, it was as if I had turned a strobe light on the spinning globe. Changes that had seemed to happen overnight now took days or weeks. Spring unfolded itself deliberately and majestically before my eyes. First came the warming of the air and the thawing of the ground; then, the first green sprouts, the opening of tiny blossoms, the arrival of birds and the awakening of reptiles; finally, the leafing of trees and the drone of insects in the air. The world, now being watched, had slowed down, and grown immeasurably larger. Nature study, the conscious seeing of the earth, is a way of slowing time and dilating space. The astronaut, traveling at great speed, finds that he ages more slowly than others; the naturalist, sitting still, may find the same is true for him.

Replacing the rock, I stand and move on through the trees. Soon I come upon a stretch of trail that winds along the crest of a ridge. In this part of the forest, the trees are old and massive; many are dying and sparsely leaved. Beams of sunlight stream through the ragged canopy, spotlighting the ground. I wind among these shimmering, slanted columns, and continue to take in the world around me, mentally ticking off the various plants and animals as they appear, occasionally pausing to make a note.

HAVING learned to see, and after spending a certain amount of time out-of-doors, you cannot help but develop a certain familiarity with the way things tend to work there. You get a feeling for the cycles, the repeated melding of one season into the next. You can predict to within a couple of days when certain flowers are likely to blossom, and know where certain species of birds will nest. Your senses grow keen: on a windy day, when the weeds are bobbing and rustling on every side, your eyes pick out the unique movement of plants caused by the passage of a snake or toad; beneath the forest

HUNTING FOR A SALAMANDER

canopy, your ears detect the muted chirps of a goldfinch flying over the treetops. Your instincts become superb: if something tells you to take a second look at a particular tree or clump of weeds you look, and find the hidden bird or the new butterfly. If you are fortunate, you learn to be receptive, and seeing, which began as a forced and active process, becomes effortless. Instead of dogging the world with your eyes, you wait as the world comes up and shows itself to you.

Once you have reached this stage, it is only a matter of time before you are asked to lead an organized nature walk, or try your hand at some environmental interpretation, as it is called. I have been asked many times, I have always given in, and I have never failed to regret it. The pattern is, by now, familiar. The group gathers; adults and children, smiling, eager to learn about the out-of-doors. I think to myself, Today it's going to work. But we begin to walk, and soon, unaccountably, I am at a loss for words. I simply do not know what to tell these people. All I can do is toss them facts, as fish are tossed to trained seals. I identify a couple of birdcalls, name a few flowers, talk about poison ivy, and the group seems pleased. Yet I am frustrated, because I am not giving them, and cannot give them, the best part of what I have learned by walking. It is a doomed enterprise, doomed from the start, because nature walking

THE OLD MARLBOROUGH ROAD

really has much less to do with simple nature than people think. The naturalist who knows every plant and every bird is like the Zen archer who, after years of practice, is able to hit the target blindfolded. The audience is astounded, but the archer simply shakes his head. The bull's-eye has little meaning for him, because he knows that the target at which he is truly aiming is himself.

The nature walk is a supremely individual endeavor. It is a private conversation between the walker and the world. One size does not fit all. When I walk I enter a private dimension, a dimension where the only things of importance are the events and objects before me; where my future is what I will find over the next rise. I step slowly and gently. I am insubstantial, ghostlike. What I see and feel has meaning only to me. I traverse a landscape partly material and partly of my own creation.

I have no need of a guide. Why should I take another's word on even the most basic fact? The particulars of this world have value only insofar as they relate to my life. I see what I see, and make up explanations as I go. So what if my conclusions are completely wrong by every recognized scientific standard? I am out not to confirm the observations of others but to experience the world for myself. Nature needs—indeed, will bear—no interpretation. My universe may legitimately differ from yours.

I EXAMINE the numerous logs and fallen limbs along the path, still searching for my salamander. What I find, oddly enough, is millipedes, lots of them. They have several traits in common: they are about two inches long, stout, and handsomely patterned in black and orange; they are all perched atop trailside logs; and they are all dead.

A quick reconnaissance reveals that the creatures are confined to a short stretch of this particular trail. I sit on a stump, take out my notebook and begin to write, getting down all the pertinent details. What was it, I am wondering, that brought them to this sunny crest to die? The little carcasses are pretty much intact. They haven't been gnawed on by anything. Were they sucked dry by spiders or assassin bugs? Perhaps it's just a seasonal thing: perhaps adults of this species mated, laid eggs, and promptly died, and I'm seeing only those who happened by chance to expire on top of logs.

As usual, I reach no conclusions, but then it is not as important to explain what I see as it is simply to see it in the first place. What I am after is not so much to understand nature as to come to an understanding with her.

I reach out, take up one of the dead creatures in my

THE OLD MARLBOROUGH ROAD

hand. In taking notice of the world I affirm that I do indeed live in a *world*. The more I walk, the more I see, the harder it becomes to fall into the fallacy of believing myself somehow separate from the rest of nature. There are times, on good days, when I seem to enter into the very fabric of creation. When my hands part the tangled vines and my shoes grip the rough surface of a tilted stone I feel an unaccountable strength surging through me. I participate in the sound and color and motion of life, and the facts (what date? what species? what temperature?) are meaningless. It is no longer important *what* I see and hear, only *that* I see and hear. In turning upon the world a receptive and disciplined eye, I move out of the realm of ordinary experience, and I feel, in the most profound sense, at home.

A breeze tosses the forest canopy, and the great beams of light begin to dance and to sway, dissolve and reform. I run a finger along the millipede's polished segments. It is difficult to explain, but unmistakably true, that my fate is somehow bound up with his; that in taking note of a millipede's death I approach, however obliquely, the mystery of my own life.

I get up from my stump, leave the millipedes and the silent hilltop behind, and walk down the path to a marshy area in a glen below. It is a promising spot. On the forest floor is a large fallen limb, a rich brown and turning to pulp. I roll it over. A centipede drops from

HUNTING FOR A SALAMANDER

beneath, writhes and quickly disappears into the loam. Sowbugs and harvestmen, startled by the light, scurry off to find dampness and shade under nearby leaves. Ants mill about seemingly without purpose. Worms slip into the earth.

Why walk? I walk, quite simply, to see the world. To see it for myself. To know what manner of world *I* inhabit.

Why the notes, the lists? They are, I suppose, a vain attempt to pick up the earth and carry it home. My notebooks are indeed rocks of a sort. They are the stone tablets on which are writ such messages as I have been permitted to bring back from the wilderness. They are a private scripture, imperfect and at times ambiguous, open to interpretation and not to be understood without some effort. Their truth, like that of most scriptures, lies as much between the lines as on them.

Why hunt for a salamander? Because, in the end, I find more than salamanders. I poke about in the weeds, stare up at the sky, investigate piles of rocks. I am alert, alive, interested. I run my fingers over the furrows in the bark of a tree. I lean to the trunk, eye it up and down from less than a foot away. I find that the eye loves to see; that there is a positive, almost physical pleasure in taking note of the details of the world. I find, too, a welcome reassurance, a renewed awareness that this world of ours is indeed one of intricacy, and order, and

THE OLD MARLBOROUGH ROAD

design. To look closely at even the most familiar place is to banish forever the possibility of taking the earth for granted. I know that nothing in this little park is rare. I also know that everything here is miraculous.

AN old, rotted slab of bark has fallen from the log and lies partially embedded in the ground. I peel it back. Beneath it, curled into a glistening little ball, is a salamander—a red-backed salamander (*Plethodon cinereus*), to be precise. Gently, I pick him up, cup him in my hand. To find him I had to know a little something about a salamander's habitat and times of activity; to identify him I had to be familiar with certain field marks. But there is nothing magical in his name, or in knowing where he lives and when he can be found there. The magic is in the way he wriggles inside my loosely closed fist. We are alive, he and I, alive and sensate and part of a community whose boundaries extend far beyond this marsh, these woods—even this earth. The tiny creature squirms; his rubbery snout pokes its way out from between my fingers, and he surveys the bright world beyond. So do I. Thanks in part to the salamander, the prospect is lovely, and limitless.

> A small soft wind blew gently from the east, a wind mild and murmurous and full of rain. It was a good omen.
>
> —Pearl S. Buck,
> *The Good Earth*

5

Rain

THE time is late spring, and I am to work the second shift at the park. When I arrive, skies are overcast. A heavy, shifting, obscuring fog, with drizzle, has settled over the countryside. Doug Cline, the morning ranger, has a look of distaste on his face. "The whole place smells like worms," he says.

He's right. And the reason is clear. Worms are everywhere. Abandoning their flooded burrows, they have crawled out onto the grass and the roads. Very few, it seems, will ever crawl back: blacktop, tires, and the soles of my shoes are smeared with a pinkish-brown, toothpaste-like substance. The air is heavy with

THE OLD MARLBOROUGH ROAD

it scent. Still, I think to myself, as Doug leaves and I climb into the patrol car, it is indeed a lovely day.

I am one of the few people I know who prefer rain to sunshine, who revel in gloom and cringe inwardly when the sun breaks through the clouds. Almost everyone can appreciate a cooling shower on a hot afternoon, but I take this sentiment to extremes. I live for those mornings that start gray and get grayer, those dripping days in spring and fall when tires hiss and gutters gurgle, when the grass squelches underfoot and bedraggled birds perch sulkily in thickets. Perhaps, as some have said, this preference of mine is mere contrariness. Perhaps, as I would like to believe, it springs from a deeper understanding of the rain.

Darkness comes early under the overcast, and the rain falls harder. With the car windows up, the atmosphere inside is stuffy, hot; the hissing of the tires and the whirring of the defroster effectively drown out the radio. My eyes are riveted on the road ahead, looking out for the telltale eye-shine of any wild creature that might be foolhardy enough to attempt a mad dash in front of an oncoming car. Some, I see, have already tried, and failed. At intervals, pulpy, varicolored masses are smeared across the asphalt. They come up suddenly and glare in the headlights. One, I can tell, was a raccoon; another, a skunk; a third defies identification.

Slowing down and peering even more intently over

RAIN

the wheel, I see other things, too, all of a sudden—gray blurs in the headlight beams, little blobs that move in a series of jerks across the road. They are alive, I realize; they are frogs, and toads, and they are on the move by the thousands. Having seen them, I feel obliged to avoid them, but the task is hopeless; there are too many. Swerving crazily to avoid one, I crush another. They make a horrible sound—a wet, explosive sound, something between splat and pop. Soon it is all I can hear. The night is rapidly losing its charm. Time for some foot patrol, I think.

I turn into the campground, and find it nearly deserted. Just as I'd figured—just as I'd hoped. The few "campers" who did not pack up and go home at the first hint of rain are sealed inside their motor homes, gathered about the friendly blue glow of their televisions like cavemen around their campfires, fending off the darkness and its imagined perils. It is their loss.

I pull into a lakeside campsite, turn off the radio and the lights, and kill the engine. The night, which is sound, envelops me. To the left, the hearty gurgle of runoff spilling into a grating; and to the right, the misty lakeshore, and the calling of frogs. Green frogs, bullfrogs, tree frogs, a few late peepers: they must fill the shallows; must cover the low branches and last year's dead cattails—countless numbers of them, unseen, in the dark fog singing, sending their collective barbaric

THE OLD MARLBOROUGH ROAD

yawp over the roof of the forest. The night is filled with their reveling, and the splash of water, and the smell of the earth, and the lake, and the worms. An ancient urge pulls me from the car and toward the lake. I walk along the shore. A couple of froglets hop by my side while their fellows bleat and trill and grunt, and the rain splashes lightly on the leaves. I sense the urgency that draws them to the water; I sense their joy.

I WAS sixteen when I first truly experienced the rain. My friend Steve Shope and I had taken up birding—with a vengeance. Every weekend would find us at some local lake or forest, earnestly examining, adjusting and strapping on our binoculars, readying our notebooks, packing three different field guides, and at last setting off along the trails, grim and resolute, alert for the passing shadow or the flash of color that might mean an addition to any one of the dozen-or-more lists we so meticulously compiled.

Early one June day, Steve and I started out for the park where, years later, I would work. The sky was overcast, and a fresh breeze exposed the undersides of the leaves. A light mist materialized and turned into a drizzle. By the time we arrived at the park, rain was falling with the lazy monotone whisper it affects when it has found a spot to its liking and is in no hurry to leave.

RAIN

Ordinarily, we would simply have turned back, spent the day at home, or at the movies, but something made us stay. Something in the wild and deserted look of the gray woods, or in the earth-smell that seeped into the car, made us decide to walk. Leaving our binoculars and our books behind (for the first time, and with great reluctance), dressed only in sneakers, jeans, and T-shirts, we set off across the parking lot and into the trees.

We were alone. The rain had chased everyone else indoors or under umbrellas, and left the forest to us. We were soon soaked, but the rain was warm and while we walked the breeze couldn't chill us. The rain sharpened the contrast between animate and inanimate nature. Logs, rocks, and old stumps looked singularly forlorn, while the leaves and the flowers seemed a shade or two brighter than usual as, swaying in the breeze, they caught what little light there was and held it close, like a silvery varnish. My glasses were streaked and useless within minutes, and when I took them off the world of leaves, twigs, drops, and tiny flowers blurred to an impressionistic landscape of green and brown mists flecked with dots of color.

We walked on and on, and felt more and more at home in the rain. Eyes nearly closed against the pattering drops, we became beasts of scent and hearing. We smelled earth, its soil and its plants; we heard water, dripping and gurgling and splashing. The rain soaked deep. It saturated our clothes, seeped into our skin, and some small essence of it filtered at last into our consciousness. We felt as much a part of the forest as the trees and the glistening stones. We were leaf mold given legs and volition; we moved with the grace and naturalness of wild things through the dripping woods.

The park was so totally deserted, so blessedly silent—even the birds and the squirrels had taken shelter—that it was as if we had slipped back to a time when only cold-blooded beasts roamed the world. We splashed and we slogged through a land that was ours completely, where no one arose to scold or to look on with disapproval. We were soaked, and covered with mud—and we didn't care. The rain stops for no one; the rain demands acceptance. And a warm rain in a green wood invites a glorious irresponsibility.

Since this first experience, I have occasionally celebrated a wet day by treating myself to a little walk. For this I have paid the price so familiar to eccentrics of every kind. Having more than once turned up in the home of family or friends looking as if I had just driven a convertible through a car wash, I have been accused, and with some feeling, of lacking sense enough to come in out of the rain. But what sort of sense is that? I ask. How great a feat is it to sprint madly for shelter? True sense, I submit, lies not in merely coming in out of the rain, but rather in knowing when it will be more profitable to come in than to stay out.

There is something to be said, for instance, for lying in a tent on a rainy night, being lulled to sleep by the drumming above, though you would rather stay awake to listen, to relish the coziness and dryness; or for staying in a cabin on a damp and drizzly day, sitting by the open window while the mountain mist, filtered through the pines, somehow gets into the steaming mug in your hands and enhances the flavor of your coffee.

I once spent a week with five friends backpacking in a red-rock canyon in the desert of northern Utah. It was one of those unspeakably rare places—spectacular, but unadvertised. It simply lay there, as it had always lain, the private wilderness of those few who by word of mouth had learned which dirt road to take from the blacktop, and which unmarked fork to take from that.

We spent the first night on the rim of the canyon, in sleeping bags on bare rock under the closest, most brilliant sky ever, amid silence so complete and cottonlike that our ears, unbelieving, filled the vacuum with tiny insect noises of their own creation—lisps and cricket chirps so authentic as to make us forget that the month was March and the temperature at freezing. In the morning we woke early, brewed and drank our coffee as the frost misted from the rocks, then shoul-

THE OLD MARLBOROUGH ROAD

dered our packs and set off, following an old trail over the edge and into the depths of the gorge. The little stream that excavated the canyon glittered in sunshine some four hundred feet below.

The canyon floor, in contrast to the rim, was warm, and green with vegetation. We followed the stream in its course, walking each day until we were tired, and then stopping to make camp.

On the fifth day we found ourselves in a canyon within a canyon—a narrow and steep-sided inner gorge choked with tumbled boulders the size of station wagons. We were worming our way through this maze of rocks when the light suddenly faded and the temperature dropped by what must have been ten degrees. We had all read up on the desert; we all knew about flash floods. In fact, at that moment, though nothing was said, each of us was acutely aware of three facts: first, it was going to be difficult to get out of this narrow channel; second, water had once coursed through here in sufficient volume to have rammed tree trunks a foot or more in diameter into crevices more than ten feet above the present level of the stream; and third, rain was beginning to fall. Simultaneously—but casually, as if for no particular reason—we picked up the pace.

Menacing, black clouds massed and churned overhead; the canyon grew darker and the cold rain sprayed and spat as it gathered strength. Under forty-pound packs, we hustled and sweated, each privately certain that his very life depended on speed. Eventually the canyon opened and the inner walls gave way to a gentle slope, which we scrambled up in search of shelter.

The rain was falling steadily by the time we found a rock overhang, sufficiently far above the stream, whose open front was screened by a row of young cottonwoods. The overhang was not deep, but by unrolling our sleeping bags in single file, parallel to the wall, each of us could keep dry.

The flood never materialized, and we savored, from our grotto, a magnificent night. Rain fell, gently and continuously, for what seemed like hours. It gathered and pooled above us, then dropped as a delicate curtain over the lip of the overhang. Its sound was all sound; its smell pervasive. We kindled a small fire from dry twigs found in our shelter, and it cast a wavering light on the shelving rocks and the windswept young trees; it caught the drops as they fell and seemed to set each one aflame. We boiled water, made coffee and tea, and cradled the steaming cups in our hands. There was little talk, much silence, and we soon crawled into our sleeping bags and nestled up against the cliff.

The fire burned low, yet still caught the drops as they fell, highlighting them, emphasizing their individuality. They fell like rubies out of the night, and shattered on the ground before me. They came from everywhere, I realized, and from all time. Picked up on a sunlit sea, they were whirled through the upper air and flung down here in the desert, only to join the stream, and then the river and finally the sea once more. The puddles I played in as a child, the primeval pool where molecules born of a lightning bolt first replicated themselves and began to grow—this is what was splashing on the rocks. All history was falling through the ruddy circle of the fire against the cliff.

The whisper of ancient times and faraway places is heard in the rain. A dinosaur thrashes in a Jurassic swamp, and the spray patters on our roofs at night. If in a distant, dusty corner of the Creator's attic there lies the carton in which was packed the new earth, then surely there must be written, in big bright letters on its lid, "Contents: One Enchanted Planet. Just Add Water."

JUST add water. The rain is earth's bath and baptism; the rain brings life to the world and to us; it beckons out of a time more remote than we would like to admit.

July 25: At the park. The heat is oppressive, the humidity palpable. Clouds gather overhead.

I walk slowly, deliberately—that's the key on these sultry afternoons. Be conscious of each step—feel the movement, heel to toe, heel to toe—feel the gravel and the grass under your shoes. Stop often, look around, do not rush.

RAIN

Goldenrod is tall, with buds but as yet no blossoms. Now is the time to watch aphid colonies—to look for lacewings, ladybirds, ants. I kneel next to the weed patch. Wilting in the heat, I poke about among the stems more out of a sense of duty (this, after all, is what I *do*) than with any real enthusiasm.

As I explore, the clouds thicken above and a few heavy drops begin to patter on the leaves and stones. Immediately, treefrogs on the nearby lakeshore begin to trill. Their chorus echoes like a hymn of praise, of thanksgiving. I straighten, feel the rain on my face. Amen.

We are dried-out creatures, living in the sun; our limbs are tense, our eyes squint. We have shrunken into ourselves and have forgotten what it is to move freely through the wilds of the world. But there is an amphibian side to our nature, if we will but recognize it. We have never completely severed our ties with the water. Listen to the rain, to the patter and the gentle splashing. It promises moonlit pools on forest floors, and calls to me to migrate overland, through the wet leaves.

A breeze springs up and the rain quickens. I feel no desire to dash for the car—deliberation is the order of the day—so I take shelter in a roadside thicket. The leaves of shrubs and saplings will hold off the worst of the shower, if it is brief, and the few drops that get by will do me no harm. The rain splashes heavily overhead and in front of me; I close my eyes to take in the sound of it, and the smell of the wet dust. I am isolated. The rain, temporarily, has stilled all movement: picnickers have taken cover, fair-weather fishermen sit in their cars, and I am undisturbed, and cozy, in my thicket. After ten minutes or so the rain stops and the sun finds a break in the clouds. The frogs continue to sing. The air is steamier than ever now, and wisps of vapor rise from the fast-drying roads.

> Most people are *on* the world, not in it.
> —John Muir

6

Bound in the Web

A FEW miles from where I grew up is a cliff that overlooks the Susquehanna River—a three-hundred-foot, beetling and fissured wall that juts from the end of a forested ridge like a stub of broken bone. From its rim a panorama of hills, fields, and villages unfolds like an exquisitely detailed model complete with flowing water, calling birds, and clanking toy trains.

During our high school years, my friend Steve and I often visited Chickies Rock, as it is called, drawn there as surely as bitterns are drawn to marshes and robins to shaded lawns. For the Rock, with its adjacent lands, was one of the vestiges of what we knew to be our natural habitat: that rarest of phenomena, the undefined place.

THE OLD MARLBOROUGH ROAD

Chickies Rock was not a park, it was not a "natural area," it was not a bird sanctuary or a wildlife refuge (at least not in any official sense). It was just a plot of ground, a corner of the earth owned but not domesticated, left pretty much to its own devices and available to those of a wandering nature. When Steve and I went there, we would stand at the very edge of the cliff, the wild updrafts whipping at our faces. We found kindred spirits in the gulls that wheeled over the bright water, and in the stunted and gnarled pines that somehow contrived to grow from fissures in the cliff's brow. Like the birds, we took from the wind something of its abandon; like the trees, we took from the rock something of its strength.

One day we approached the cliff from the bottom and, without really having planned it, found ourselves working our way upward through a jagged cleft in its face. We had no climbing gear—no ropes or pitons or proper boots—but we inched along, like slugs or creeping insects. We grabbed at roots and tiny points of rock; we wriggled through a rich growth of poison ivy; we dismissed, as only the young can, the thought of twisted limbs, shattered skulls, pooling blood—and we made it to the top. Standing there, we felt not conquest but camaraderie: we had not beaten the cliff; we had simply come to know it better.

* * *

CHICKIES Rock, as Steve and I knew it, was too good to last forever. The Rock was no secret—it was well trod by naturalists, climbers, lovers, and the local rowdies—but, however crowded, it always felt like a private place. There were no restrooms, or signs, or picnic tables, or self-guiding trails—none of the trappings that define and delimit an area, mark it off as having been prepared, by others, for some specific purpose. Rather, undefined, it was whatever we wanted it to be. It belonged not to The People, but to the people, the individuals, who loved it. It was only a matter of time, we were sure, until somebody else—somebody in *charge*—got wind of the treasures hidden there. And then, of course, everything would be ruined.

In due course, the inevitable came to pass and the Rock and adjacent lands came into the protective custody of the county's department of parks and recreation. Parking lots and picnic areas were built, and the process was begun of transforming what had been just a place into a "recreational resource." At first, this development was carried out in areas I had never really frequented; the sacred core—the cliff itself, and the woods and trails immediately around it—was left untouched.

But not for long. For despite the Rock's new and official status, and despite an increase in its regulation, students and other revelers continued to view it as the ideal party spot. They came after dark to set up camp in the woods, and they carried their six-packs to the edge of the cliff in the moonlight, with predictable consequences. In the end, it was chemistry and physics that changed Chickies Rock forever: the chemistry of ethyl alcohol as it affects the human brain, and the physics of gravity and acceleration as they act on a human body plunging three hundred feet onto unforgiving stone.

Accidents at the Rock were nothing new, of course. But what were seen, in years past, simply as personal tragedies, suddenly came to be viewed as matters of public concern. Grumblings were heard, calls for action.

Amazingly, some of these grumblings were directed not toward those who made it a practice to get drunk at the edge of a precipice, but rather toward the proprietors of what was seen as a dangerous piece of real estate. Letters to the local papers spoke of the need to safeguard the public from a clear and present danger. By implication, the cliff itself was at fault. It was as if it had snuck up on some poor innocents while their backs were turned, and withdrawn itself from beneath their feet with malice aforethought.

The cries of indignation grew louder, and there were rumors of change in the wind. According to the papers, warning signs were being posted at the edge of the cliff, and their message would soon be given substance in the form of a fence.

And that, of course, would be that. The transformation would be complete. I set out for one last look.

IT'S just a short walk from the roadside to the cliff. The trail rises steeply at first through a sparse and spindly wood. Here the trees, long in battle with strangling vines, are bent and corkscrewed as if grabbed and wrung out by passing giants. Emerging from the woods, the path levels off and traverses the flank of the ridge. To the right rises the rocky slope, and to the left is a wild hedge of brambles and multiflora rose. Beyond and below this hedge lies a sun-beaten bowl of impenetrable scrub that gives way at its far edge to another patch of scraggly, vine-smothered forest. Along this open stretch of trail indigo buntings sing and the smell of honeysuckle is strong. In depressions along the path, water from recent rains gathers in green-rimmed puddles, bright with sky and tumbling clouds. After a few hundred yards, the trail turns left and descends a gentle slope through woods to the final drop-off of the Rock itself.

It was not yet noon, and I met only two or three people along the path. As I walked down among the trees, it seemed almost like the old days. Then, as the far shore of the river first came into view over the cliff

THE OLD MARLBOROUGH ROAD

ahead, I saw the signs, a whole row of signs. They stood some ten to twenty feet from the brink of the Rock, extending to the right and left as far as I could see along the rim, and they warned:

**DANGER!
DO NOT GO
BEYOND THIS POINT.**
Carelessness at this overlook has resulted in serious injury or death. Unauthorized persons found beyond this point will be cited for criminal trespass per PA Crimes Code, subsection 3503.
Throwing stones or other objects from this overlook is prohibited. Please exercise extreme caution in this area.

LANCASTER COUNTY
DEPARTMENT OF PARKS & RECREATION

My first response was to critique the phrasing. Serious injury *or* death? Didn't they know which? It was only after allowing myself a chuckle that I began to grow angry. It seemed a judgment had been passed upon this place. The Rock had been tried and found guilty of being a public nuisance, a hazard, a thing to be avoided and feared, and this phalanx of signs was the sentence handed down. It was the cliff's badge of shame, its Scarlet Letter. I felt as though a friend were being publicly humiliated in front of me—and there was nothing I could do about it. Frustrated, I moved back into the shadow of the trees and sat down on a rock to think.

Sitting there, I was distracted from the problem at hand by a movement at the periphery of my vision. Turning, I saw a pool of dark water that had gathered in a hollow between two stones, and in it a small brown moth. The moth had fallen upside-down, and was held fast to the surface film by his wings. He struggled desperately to escape, but his violent thrashing served only to propel him, like a tiny motorboat, from one cliff-bound shore to the other. It was clear that, barring outside assistance, his situation was hopeless.

Instinctively, I reached out to the insect—but then I stopped short, and slowly sat back down. The business of the Rock and the signs had thrown me into a cold and perverse frame of mind. Lips pursed, head cocked to one side, finger tapping chin, I watched the moth and I pondered.

What does it really mean, I asked myself, to pull a moth from a puddle? For instance, are there not hungry birds to which this struggling insect would represent a meal? And are there not scavengers that could feed on the moth's carcass, and flies that could deposit their eggs on it? To save one life, then, might lead to the loss of others. And exactly what sort of beast am I dealing with, anyway? Could he be a gypsy moth? The size and color are right, but I'm not sure. And if he is, how many of his kind will he sire, and how many trees, or acres of trees, will they defoliate? And what about the future of moths as a species? Would I not be circumventing natural selection, derailing the course of evolution, by saving an individual so clumsy, so obviously unfit to survive?

On the other hand, I thought, here before me, in the present, is a living thing that is about to die. Of all the world's creatures, only a human being would stop to help him; and of all the world's humans I sit before him now.

The more I thought, the larger the little moth loomed. It became evident that his living or dying would send ripples through time just as surely as his struggles, almost half-hearted by now, sent ripples across the surface of the puddle. Good lord, I thought, the fate of the world could hinge on this decision.

Whether or not to reach into a puddle and pull out a bug was snowballing into an ethical question of enormous magnitude, and I needed help to resolve it. I looked to memory for guidance. I thought of all the nature programs I had seen on television. One had shown me the slow and agonizing death of a tortoise that had somehow been flipped onto its back and was unable to right itself. As the creature baked in the heat of a tropical sun, the camera caught every pathetic and futile kick of its legs, right down to the last. Another show had featured footage of a seal pup that had slipped through a hole in the arctic ice and was struggling to save itself from drowning. The pup's cries of terror echoed in my ears for a long time afterward. Yet another had treated the audience to a close-up shot of a den of young animals—were they skunks? I couldn't remember—whose mother had been killed by a predator. As the faces of the helpless, orphaned creatures filled the screen, the narrator intoned matter-of-factly that, bereft of their mother, the babies were going to die. Nobody had reached out to the tortoise, the seal pup, the young skunks. That, the films had seemed to imply, would have been unrealistic.

I thought, too, about the biologist who had once described to me how he had encountered a small shark left stranded on a beach by the receding tide; how he had struggled with conflicting impulses; and how, in the end, he had decided to "let nature take its course," that is, to let the shark suffocate in the sand.

I had not agreed with the biologist's decision, just as I had not agreed with the decisions made by those filmmakers. Something in their reasoning had bothered me. They had left something out, I thought, though I could not, at the time, have said what it was.

But now I knew. I saw the fallacy in their logic—the fallacy that had brought about the deaths of hapless

THE OLD MARLBOROUGH ROAD

sharks and tortoises—the same fallacy that would lead to the posting of signs and the building of a fence at the edge of Chickies Rock. It was an error in self-perception, an error having its roots in the way people view themselves in relation to the world.

Simply put, we have not yet decided what we are. One moment, the mirror on the wall reflects a being divinely shaped to assume rightful dominion over nature. The next it reveals an evolutionary freak who has seized power and threatens to destroy the world. All we know is that we feel somehow separated from the world around us, that we seem to hover about on the edge of nature, living on the earth but not a part of it in the same way as, say, an acorn, or a vulture, or a chunk of quartz. This feeling runs deep, even in those who most loudly profess their oneness with nature. I find it in myself. I speak of the natural world, and I speak of man. The distinction is sharp, the separation complete. To be human, it follows, is to be "unnatural."

Without being able to say precisely why, we find ourselves outside the world looking in, waiting for someone to tell us what to do. As a result, we move through life hesitantly and bewildered; we act, when confronted by nature, like a ten-year-old boy in the presence of a pretty girl, unsure whether to kiss her, to chase her with a frog, or to run for his life.

One half of the psyche is ashamed of itself and our

species. We are incessantly tinkering with the world, that half says, sticking our fingers into everything and messing up what would be, if not for our presence, perfection. To this half, the nature of man's interference is immaterial because any interference is a disruption. Reaching out one's hand to an animal in trouble is as wrong as bulldozing a forest.

The other half sees itself in a battle for the domination of the planet. Nature, to it, is a rival to be overcome; it is the obstacle that stands between where humanity is and where it is destined to be. We've got to get nature, it says, before nature gets us. This half has grown accustomed to, and takes as a matter of course, automobile accidents, chemical spills, assaults, murders, and warfare; yet it cannot abide the thought of being eaten by a grizzly, being bitten by a snake, or falling from a cliff. In its view the phrase "death by natural causes" has come to mean, paradoxically, being old and out of one's senses, and taking one's final breath while strapped to a hospital bed and wired to a roomful of machines.

In the back of our minds the two halves of us conjure up their images, one juxtaposed against the other: we hear the rumble of killer earthquakes, and smell the stench of our own pollution; we feel the leopard's fangs at our throats, and see the last wild passenger pigeon crash dead to the ground. We may be Creation's mas-

THE OLD MARLBOROUGH ROAD

terpiece, or we may be evolution's mistake. Uncertain, we waver between one extreme and the other; hover somewhere between the two poles; keeping our distance from the world lest there be trouble; hoping not to step on a bee because (1) it may sting us, and (2) we may kill it.

It is only our unique ability to trace the chain of cause and effect that makes everything seem so complicated; it is only our habit of continually looking toward the future that paralyzes us here in the present. In reality the situation in which we find ourselves is the same one that has faced every animate creature since the first amoeba wriggled in the primeval sea. We affect the world and are affected by it. There is a simple phrase for this predicament: it's called being alive.

The truth is that no matter how often we may feel alienated and estranged from the world around us we are nonetheless a part of nature, inextricably bound in the web. When we move, the whole web vibrates, and when anything else in the universe moves we feel it. We act and are acted upon; we harm and are harmed; we help and are helped.

Our least actions can have unforeseeable ramifications. Once, on a winter day, I came across a discarded bottle lying in the snow. The snow had melted for half an inch around the bottle's perimeter and inside, a brilliant green in contrast to the bare ground and the leafless trees, was a clump of moss. It had found itself a hothouse in a cold world and was flourishing there, thanks to the seemingly thoughtless act of a litterbug.

When we take a step, when we break a spiderweb, when we mow the grass, when we idly turn over a stone, when we burn a dry branch—when we do these things we alter the shape and direction of the world. Just to get up in the morning is to set in motion ripples that change the course of history. We cannot extrapolate into the indefinite future; we can no more predict the ultimate consequences of our actions than we can predict where the next meteorite will fall to the ground. Nor, beyond taking reasonable care not to cause unnecessary hurt or take unnecessary chances, can we dwell upon the upshot of all that we do. Francis Thompson, in an oft-quoted verse, wrote that one can't stir a flower without troubling a star. This is undoubtedly so, yet in our day-to-day lives it is enough to realize that in stirring a flower we can trouble the *flower*. To understand this, and to act accordingly, is all that we can ask of ourselves.

I reached down and carefully scooped the moth from the puddle, cupping him in my palm. Then I got up, strolled past the signs and stood with my toes hanging out over the edge of nothing. The river far below was

THE OLD MARLBOROUGH ROAD

high, muddy. Debris swept past on its surface—bits of vegetation, broken branches, boards. Over the breeze, I could hear the deep, insistent murmur of the water, an ancient and comforting sound. The wind tugged at my hair; the gnarled pines whispered softly; the Rock was firm, and supporting, and trustworthy beneath my feet.

I opened my hand. In the end, it was a decision of the moment, a decision of the heart. The moth would have a second chance. We all deserve a second chance. I am a human being, and unpredictable by nature, but I am woven hard into the fabric of the world around me, and whatever curiosity, whatever compassion I may possess—though they are subject to errors in judgment—are as fundamentally natural as the lion's claws and the flower's scent.

The moth sat on my palm for a few moments as its wings dried, then took off on the breeze. I watched as it caught an updraft and fluttered away, finally vanishing from sight high over the river toward the far hills.

Epilogue. One year later: At the Rock, all has come to pass as foretold. Visitors, once able to wander with the wind along the very brink of the cliff, are now shepherded into a couple of official, fenced, scenic overlooks. The fences, I must admit, are quite tasteful. Stonework columns joined by pairs of varnished logs, they present no real barrier. They are merely symbolic, one assumes, designed to the specifications of attorneys and insurers. Nothing of real significance has been changed here: the wind is as capricious as it ever was; the sky is just as blue. Still, in a way never to be explained to those who do not already understand, the view has been diminished immeasurably.

> 'God's humblest, they!' I muse. Yet why?
> They know Earth-secrets that know not I.
> —Thomas Hardy,
> "An August Midnight"

7

The Creatures of Contradiction

APRIL. At the park, all is calm and unhurried. The parking lots are empty, and lake, wood, and lawn lie still and shimmering in the sunlight that follows a morning rain. But there is a sound in the air, a soft, sustained and musical humming that might be the sound of the awakening earth itself. It comes from a lone tree growing at the water's edge, a weeping willow, fuzzy with flowers. Approaching, I realize that the tree is covered with bees, hundreds of them, hovering about and clambering over the new, green blossoms, eager for the first nectar of the season. This is spring.

Just across a marshy inlet, growing from a rock on

THE OLD MARLBOROUGH ROAD

the shore, is another willow, a shrub. It has no flowers—looks nearly dead, in fact. Its branches are frayed as if hacked by a rusty saw, and beneath the tattered bark I can see the tunnels of insects. A larva of some sort, white and legless, writhes within one of them. There must have been many more. I see it in my mind, as if in time-lapse—scores of maggots bursting out all at once, shattering the wood, shredding the bark—and I shudder. This, too, is spring. For latent in spring, in its buds and blossoms, its nectars and perfumes, its sproutings, unfoldings, and gentle becomings, is all the appalling rankness of summer: the impenetrable thicket; the scum-covered pool; the stench of decay—and the reawakening of the bugs.

Bugs. Butterflies on brilliant flowers, and bluebottles raised on piles of filth; ladybirds devouring pests, and locusts destroying crops; fireflies over a twilit meadow, and roaches massed on a tenement wall.

Bugs. Bright, metallic armor, and pale, pulsating flesh; the gaudy fliers, and the hidden crawlers; the makers of honey, and the carriers of disease.

Bugs. The subjects of poetry. The stuff of nightmares. The creatures of contradiction.

I HAVE gone through a number of stages in my nature-study career. In my teens I was a birder. Since

THE CREATURES OF CONTRADICTION

then I have passed through flower phases and tree phases, and have spent weeks at a stretch absorbed in fish, frogs, mushrooms, and snakes. But in my earliest childhood I studied what I could get my hands on, the butterflies, grasshoppers, and caterpillars I could find in the backyard and the schoolyard. I began with bugs, and to bugs I always return. Bugs never disappoint. They always offer something to see, always provide food for thought, no matter how otherwise boring and uneventful the day. Bugs do not have to be cajoled or stalked: they come to you. Bugs pose countless questions. They promise even the amateur (if he or she is interested in such things) a real chance of adding to our shared knowledge. Bugs are a wide-open field. There are more than enough of them to go around, and they are good for a lifetime of study.*

BUGS as a group have an image problem. Despite their ubiquity, despite their diversity, despite their ofttimes breathtaking beauty, they get little respect from

---

*I am perfectly aware that the term "bug" properly applies only to insects of the order Hemiptera, the so-called "true bugs." I use it here in a broad sense to refer to insects, spiders, harvestmen, sowbugs, centipedes, millipedes, ticks, and those other relatively small, multilegged creatures that people tend to swat, step on, run away from, or scream at the sight of. I use it in this way because it is convenient and, to a certain extent, because it irritates the hell out of specialists in the field—I do so love to be difficult.

the average human being. People who, out of compassion, cannot bring themselves to eat meat or wear leather will step on a spider without a moment's thought, and feel virtuous. Bugs, it seems, somehow fall short of being considered legitimate animals.

In part I think the problem is one of simple size, or rather the lack thereof. We fail to respect the bug because we seldom really see him. Ours is a life of almost constant motion. Striding purposefully from place to place, riding and driving at ever faster speeds, we are forced to devote most of our attention to simple navigation: to getting in the proper lane, finding the right building, tiptoeing around the dog droppings in our path. Bugs, while occasionally annoying, are simply too small to impede seriously our immediate progress through the world, and so they are overlooked. We see the sidewalk, but not the anthills in the sidewalk cracks. We see the billboard, but not the tufts of grass at the base of the billboard, and there are whole worlds in those tufts of grass.

Not too long ago I lived in the city. At the rear of my home lay a typical urban yard, about fifty feet long by fifteen feet wide. It was essentially a swath of crabgrass and dandelions, with a handful of ornamental shrubs and a couple of small trees. On one side, however, next to a wire fence, was a narrow strip of uncultivated soil

THE CREATURES OF CONTRADICTION

that was home each summer to whatever vagrant weeds happened to establish themselves. This small patch of bindweed, dead-nettle, nightshade, and goldenrod was my day-to-day link to the world beyond the asphalt. Whenever I got sick of my urban existence—which was often—I had only to step out into the yard, kneel down, and peer into the weeds. I would stare into the greenery until I forgot the claustrophobic proximity of neighboring houses and the rush of traffic in the street. I explored the weeds as one would explore a forest. I found aphids, honeybees, bumblebees, ants, jumping spiders, ladybirds, lacewings, longhorned beetles, and hosts of others, and I watched them as country people might watch birds or deer.

When you take the time to watch bugs, you are apt to have a startling revelation: bugs walk and bugs run; they eat and they excrete; they fight, they couple, they give birth, they grow and they die. In other words, they do what we do; they need what we need. They are, quite simply, alive—as alive as you and I. Moreover, it soon becomes obvious that in terms of drama, in terms of adventure, their lives put our own to shame.

Consider the ordinary scene before me: ants are dragging the carcass of a caterpillar across a broken expanse of concrete to their nest. Imagine it from their perspective; put it in human scale. You are an ant, say

THE OLD MARLBOROUGH ROAD

six feet long, two feet high. You and a few of your buddies are dragging a caterpillar nearly a hundred feet long across a broken and pitted landscape almost two miles to your burrow. Straining at the huge, horned-and-pimpled beast, you slip and find yourself at the brink of a gaping pit in the earth, a conical crater forty-eight feet across and thirty-six feet deep. The soil at your feet gives way, and you begin to fall, rousing the maker of the pit, the ant lion larva that rests buried at the bottom. With convulsive jerks of its huge body, the larva creates a shower of boulders, each the size of your head. They soar twelve feet over the rim of the pit and come crashing down on you, driving you toward the bottom. There, rising from the sand, the last thing you'll ever see, is a pair of jaws. They are sickle-shaped and gleaming. They are nine feet long. And they are coming together to impale you through back and chest, to siphon out your innards and leave of you a dry and shriveled husk that will be tossed back out of the pit like so much garbage.

Or consider the following: I stop my car one night at the edge of a lake. The headlights illuminate a section of the shoreline, and I see that the surface of the water is covered with tiny mayflies—subadults, I assume, newly hatched and trying to make it to shore. They are trying to fly, but are being blown down by a rushing wind.

THE CREATURES OF CONTRADICTION

Individuals progress in a series of hops across the lake's surface. Some of them disappear before my eyes as small fish flash briefly in the headlights. Put yourself in their place. Having spent the first year or two of your life underwater, as a flattened nymph creeping about the quiet silts, one night you attain your adult form and rise to the surface of a dark and storm-tossed sea. You test your new wings, launch into flight, but you are slapped back down onto the water. You perch there on the surface film, as the wind howls above. You ride the flanks of thirty-foot waves, gathering your strength for another attempt at flight, while on every side fish the size of submarines erupt from the water and then dive in a cloud of spray, leaving the surface each second more empty of your fellows, leaving you more and more alone under the wind and the glittering stars.

Or this: I am strolling along a country lane on a hot, humid afternoon. A caterpillar crosses the road in front of me. The little guy is positively running, and no wonder, given the probable temperature of the blacktop. Look at it from his angle. In your travels you have reached the edge of nothingness. The green and benevolent world has simply come to an end, and before you is an uncharted void, a shimmering plain, black and scorching, barren and, for all you can tell, endless. It is not in your nature to turn back. A faint smell, perhaps only a faint hope, leads you on. Belly to the burning ground, you strike out in search of new, more congenial lands.

THE bug as an individual is an intriguing, comprehensible, even likeable creature once you get to know him. You can respect, even empathize with, the caterpillar crossing the road. Bugs in general, and bugs in numbers, are quite another matter. We often hear how many bugs there are in the world, and what fecund little beasts they are. Many of us have read how in the course of a summer, in the absence of population controls, a single pair of flies, producing offspring that reproduce in turn, could blanket the earth with insects to a depth of some fifty feet. But it's difficult to get a firm conceptual grip on such numbers. To really understand the terrible potential latent in the gonads of a bug you must see a few million bugs together, in the same place, going about their business utterly oblivious to your presence.

When you think of bugs in great numbers you naturally think of South America and its legions of army ants, or North Africa and its swarms of locusts. We in the temperate regions are for the most part deprived of such sights. However, in the spring of 1987 people in the northeastern United States were treated to a taste of the tropics.

That year, as the weather began to warm, an unseen migration was set in motion. As we went about our

THE CREATURES OF CONTRADICTION

daily business—as we worked and played, ate and slept—pale creatures began to stir beneath uncounted acres of woodland. White nymphs, grublike but with powerful, clawed limbs, left the rootlets from which for nearly two decades they had been sucking sap, and began to dig toward the surface, under our feet, by the billions. The periodical cicadas were reaching maturity.

BY April 26 it is evident that something is stirring within the earth. At Pinchot Park, the forest floor is raised into odd bumps and tubercles. Two inches tall, an inch wide, they look like huge worm castings, but are connected to the ground, not lying upon it. It is as if the earth had broken out in pimples, or the ground had boiled during the night, and the muddy bubbles had congealed in place. Somewhat hesitantly, I break off one of the earthen lumps. Within it is a hole, perfectly smooth, as if someone had poked a finger into a ball of wet clay; beneath it, a hole of the same diameter leads down into the earth. Later, at home, I read that these structures are the work of cicada nymphs. According to my book they are called "turrets," and their purpose is not well understood.

May 15: I walk in the park today, and in flipping a large rock I get my first glimpse of the young cicadas. It is an unsettling glimpse. The creatures are pale and soft; startled, they back down quickly into their burrows.

THE OLD MARLBOROUGH ROAD

They are waiting for a signal I will not perceive; they are not yet ready for the light.

May 28: Sometime during the past week the emergence began, and now the cicadas are among us in force. Discarded skins, amber husks slit down the back, cling to the grass and the trunks of trees. The newly emerged adults, each more than an inch long, are black with orange eyes and wings. They are moving toward the treetops, and the woods resound with their calling. I make out three different sounds: (1) a series of clicks followed by a wheezy snore of one to two seconds; (2) a series of stacatto hisses, like maracas shaken three or four times in quick succession; and (3) a half-heard whine coming from everywhere at once and no place in particular—as constant as a ringing in the ears, and always heard at a distance. There are many individuals on low plants and shrubs, but the singing comes only from the treetops. Perhaps those at lower levels have only recently emerged; perhaps they are too preoccupied with climbing to sing, too keenly aware of the ants and the other predators eager to catch them before their new bodies harden.

I find one individual fresh from its nymphal skin, which is anchored to the underside of a leaf. The creature has crumpled wings and jet black eyes; it is a hideous, ghostly white.

The insects seem to have emerged in pockets, corresponding perhaps to groves of favored trees, or to segments of the lakeshore that have escaped sudden flooding over the past seventeen years. In some places the ground is shot full of holes; in others, few of the insects are found.

Along Alpine Trail the low plants are covered with shells and insects to the extent that I am unsettled by the sight. Scores are seen at one glance. The treetops shrill with song—a low snore that swells at intervals to a great communal exhalation. Looking closely, I see movement—creatures crawling in the weeds and up the tree trunks, a migration steady and inexorable. Looking even more closely I draw back, sickened. A cicada crawls up a stem, and I pick it up to check its gender. Turning it over, I find that it has no abdomen, just an empty, gaping socket at the end of its thorax. Half of its body is gone, and yet the creature climbs. What beast hidden in the weeds feasts so delicately, plucking insect abdomens as you would pluck apples from a tree, leaving the rest untouched? Shrews, perhaps? I don't know. Looking deeper into the greenery I find another cicada in the same fix, and yet another. All still move, still climb. They cannot sing, they cannot mate; they are for all practical purposes dead; and yet they climb, climb in obedience to ancient law.

June 7: The emergence is at its peak. How many cicadas are there in Pinchot Park? On one bush, cover-

THE OLD MARLBOROUGH ROAD

ing about a hundred square feet of ground, I count twenty or more (the population is much denser in the treetops). To be conservative let's make it one individual for every ten square feet—a gross underestimate if anything. That's eight million insects in two thousand acres. The actual number is probably ten times (and possibly a hundred times) greater.

By two o'clock temperatures have climbed into the eighties, and the cicadas, reveling in the warmth, are going about their business with an urgency and intensity that are almost palpable. Their screams are deafening, and inescapable; I am in an ocean of sound. The bushes and the grasses are a moving mosaic in black and orange; the air is filled with ungainly flying forms. Jostle a branch and a dozen bumbling bodies scatter, squawking. They sail out, double back, bump into my chest, my face. I step on them unintentionally; I cannot help it. The sound is unending, and maddening. Individuals shake like maracas. They snore, they emit an eerie, disembodied whine, "brrrr. . . . owwwwl." Everywhere, bugs creep and climb and fly and drop and clatter and squawk. Everything within my field of view is fluid and shifting with the massed movements of cicadas. Birds go unnoticed, flowers go unnoticed. The world belongs to the bugs, and no one disputes their ownership today. Walking along a narrow path I feel as if I am wading through a waist-deep mass of squirming insects. One minute I thrill to the sight, and the next I recoil with an involuntary shudder. Bugs are like that.

THERE is a species of ant in North Africa whose queen, once inseminated, loiters about the entrance to another species' nest until she is captured and dragged inside. Once inside, the invader queen gets loose and mingles for a time among the host workers. Eventually, in the words of entomologist Edward O. Wilson, "she settles down for good on the back of the host queen and begins the one act for which she is uniquely specialized: slowly cutting off the head of her victim." Over the course of several hours, the invader queen saws with her mandibles at the spindly neck of her rival. The grizzly murder done, the invader takes command of the nest and commandeers its workers to rear her own offspring.

Bugs rest squarely on the fine line that separates wonder from revulsion. Seen closely, observed carefully, they attract, but their attraction is a morbid one. They are compelling, but in the same way that twisted steel on the highway, chalk outlines on the floor, and crusted pools of blood are compelling. The naturalist is drawn to them in spite of himself.

Bugs, it would seem, have no inhibitions. If there is a thing that can be done, there is a bug to do it. As Annie Dillard puts it: "Fish gotta swim and bird gotta fly;

insects, it seems, gotta do one horrible thing after another."

The sobering truth is that the bugs are the most numerous and successful creatures on earth. Their lives, as horrible as we may think them to be, are nonetheless more typical than our own. They are the rule, and we, the exception. We may look upon their behavior with outrage or with disgust, but a part of us recognizes that the bugs as a group represent a force not to be taken lightly. A tiny voice in the back of our collective mind whispers that the bugs outnumber and out-reproduce us; that they are supremely adaptable; that they can resist our most potent pesticides, and survive our nuclear blasts; that they can waft high in the atmosphere without plane or oxygen, and come down softly and unperturbed; that they can live for incredibly long periods without food or water; that they have (as far as we know) no conscience; and that they feel no remorse. The voice warns us that were we, for whatever reason, to lose our grip on this world, we would find, there in the blackened grass, ready to take our place, the bugs.

No doubt many people would be frightened, or depressed, or at the very least humiliated at the thought of man's being supplanted by bugs. But I find in this prospect a measure of reassurance, even a certain justice. Whenever I see a new development being built, I think of bugs; whenever I see waste poured onto the land or into the sea, I think of bugs; whenever I fear for the survival of life apart from man, I think of bugs.

IN 1883, at the age of sixty, Jean Henri Fabre settled into a house on the outskirts of Serignan, France, and began doing what he had always really wanted to do: he studied bugs. For three decades he explored the two acres of "barren, sunscorched . . . land, favored by thistles and wasps and bees," enclosed by his walls. On this humble stage Fabre observed flies, grasshoppers, and caterpillars; bees, wasps, beetles, and spiders; and he wrote of them all with perception, wit, and empathy. His ten-volume *Souvenirs Entomologiques* remains one of the most informative and entertaining accounts of bug behavior.

Fabre was a confirmed experimentalist. Endlessly, he tormented his arachnid and insect neighbors, devised ever new ways to test their abilities and mess with their minds. If a wasp habitually grasped her prey by a certain one of its limbs, Fabre would snip off that particular limb when the wasp wasn't looking, and sit back to await developments. If the same wasp left her prey unguarded as she inspected her nest burrow, Fabre simply stole the prey and waited to see how the wasp would react. Fabre came to the conclusion that bugs are essentially automatons, mere slaves to instincts that, although useful within a given range of conditions, are

THE OLD MARLBOROUGH ROAD

positive hindrances when circumstances change. "I am . . . familiar," he wrote, "with the abysmal stupidity of insects . . . whenever the least accident occurs."

This, then, is our final shot at the bug. He may conquer the world, we think to ourselves, but he'll never be able to revel in his conquest—because he's stupid. We like to think of instinct as the primitive ancestor or at best the ugly stepsister of true intelligence, an inferior strategy for survival. The fact we cannot argue away, however, is that the strategy works.

Instinct is, in a very real sense, a form of simple trust—trust in the earth, and in the workings of the earth. In the world apart from man, "accidents" that could threaten the survival of a species happen comparatively rarely (and bugs, it should be noted, never cause their *own*). Worldwide change on the order of ice ages and continental drift is usually gradual. On the other hand, catastrophic change (fires, floods, volcanic eruptions and the like) may happen quickly but tends to be local. The bugs are merely playing the percentages. They are betting on a tremendous fecundity, on dispersal over a wide range, on a rapid turnover of generations and thus a rapid rate of evolution, and on a set of largely preprogrammed behaviors that kick in almost automatically under the right conditions. They seem to have bet wisely. As a group, the bugs have

THE CREATURES OF CONTRADICTION

proven themselves able to handle pretty much anything the natural world can dish out. Even confronted by such a beast as man, and by consequent changes of unprecedented rapidity and scope, bugs seem to be holding their own. Ask any cockroach.

We simply cannot dismiss the bug. We cannot call weak the ant that hauls a caterpillar; we cannot call fainthearted the cicada that, half-eaten, continues to climb; we cannot call unnatural lifestyles more common and widespread than our own; we cannot call inferior the most successful creatures on earth.

As the years pass I am less and less convinced that we can make any unequivocal statements about nature at all. We can no more read the mind of an insect than detect the subtle odors—the sex attractants, the alarm pheromones, the trail markers—by which he steers through his world. Don't judge a bug until you've walked a mile in his exoskeleton. It is possible, I think, that there is more going on behind those gleaming compound eyes than even the most freethinking naturalist would care to admit publicly; that those little lives are fuller and more rounded than they appear; that perhaps, just perhaps, the bugs are tied into the fabric of the world in ways we cannot begin to imagine.

Take, for example, an incident that occurred halfway around the world, on an evening in late summer in Sierra Leone, West Africa.

THE OLD MARLBOROUGH ROAD

I was a Peace Corps volunteer then, living in a mud-walled house that stood in the shadow of the bush at the edge of the up-country village of Makali. That evening I was out on my veranda, smoking my pipe, when from the east came a muffled rumbling, as of the toppling of distant towers. Soon the sky overhead grew dark, and a cool breeze set the leaves to dancing in the eerie, orange light of the sunset. The thunder grew louder, the wind grew stronger, and I could hear the hiss of the approaching rain over the roofs and trees to the east.

Villagers dashed for shelter as a few preliminary drops smacked the dust on the road, and a moment later all hell broke loose as the skies poured forth a torrent of what might have been molten lead: The driving rain was picked up and hurled horizontally, or swirled around and brought crashing down on the metal roof with the force of breakers on a rocky shore. Strands of blue fire snaked out across the sky, and the little yard behind my house seemed lit by some huge, celestial strobe light. Each puff from my pipe became for a second a pulsing nebula, and then rushed off before the wind. Thunder from every quadrant echoed and overlapped in a terrible and incessant roar. A drenching mist filled the veranda, and through it I paced, and worried, and prayed for the preservation of my walls.

The storm, though violent, was short-lived, and after

THE CREATURES OF CONTRADICTION

thirty minutes the thunder had grown faint over the western hills, and distant lightning now and again flickered harmlessly in the twilight.

The sky between the clouds was a dark and bottomless purple, and I stood on the veranda listening to the drops as they slipped from the leaves and the roof and pattered softly to the soggy ground. A dozen or so fireflies began to glow in my little lawn, some shining steadily with a faint greenish light, others dimly blinking beneath the grassblades. All was tranquil, and quiet, and beautiful.

Suddenly, there came a flash of sheet lightning, brilliant against the broken clouds. As the yard was cast into darkness once again and the distant thunder began to rumble, the entire expanse at my feet lit up with fireflies—fifty or more, glowing with redoubled brightness. I watched, entranced, as half of them flashed in perfect unison—once, twice, three times—then all faded and were gone.

It is entirely a judgment call, of course, and the evidence, as always, is inconclusive, but how we regard the creatures with which we share this planet will depend upon how we answer this question: Was this merely an example of instinct gone amiss, or was I privy to a conversation between some insects and the sky?

> We've looked and looked,
> but after all where are we?
> —Robert Frost,
> "The Star-Splitter"

8

The Hidden Wilderness

A NUMBER of years ago, I humored that part of me that had always wanted to be a wildlife biologist by volunteering to monitor the nesting of bluebirds at the state park where I worked. Each Saturday I make the rounds of a dozen or so boxes, carrying a clipboard and taking the census. At each box I would knock politely on the wall. Once the adult occupant, if any, had flown, I would take out my screwdriver, remove the roof and peer inside. There I would find eggs, or young, or, occasionally, the defiant stare of a particularly gritty mother bluebird. I would note my findings, replace the roof and proceed to the next box. I felt scientific, and

THE OLD MARLBOROUGH ROAD

useful, and I enjoyed my work—except when it came to the last two boxes. Like the rest, they were situated in old, regenerating fields, but to get to them I had to park at the dead end of a little-used road and follow a telephone company right-of-way, a brushy corridor about forty feet wide that angled up the side of a wooded hill. In the spring this clearing was easy to negotiate, but as the nesting season drew on, new growth—most of it briers, wild raspberries, multiflora rose, and other thorny plants—made walking more and more difficult. By the middle of June I dreaded this part of my rounds; by the end of summer the clearing was simply impassable.

June 21: At two o'clock it is hot, unseasonably hot. The sky is utterly cloudless, and there is no hint of a breeze. I am parked at the end of the road, looking up through the right-of-way at the tangled greenery that chokes it to a depth of some four feet. The boxes really should be checked, but I don't want to go up there. In fact, I'm inclined to turn around and head directly for the nearest air-conditioned bar.

In search of guidance, I mentally review the many episodes of *Wild Kingdom* I watched in my younger days. What would Marlin Perkins have done in a situation like this?

He'd have sent his old buddy Jim up there, and watched from the car, that's what.

Fresh out of loyal sidekicks, I have little choice but to go in myself. Nobody said being a scientist would be easy.

The path, what there is of it, is all but overgrown—difficult to follow, difficult even to find. Briers reach in from both sides; their thorns have grown stout, wickedly hooked, menacing. I step on top of the low plants, pinning them to the ground, and grasp the high ones by the leaves, with two fingers, gently lifting them up as I pass under. Easy, easy, I tell myself, you are in complete control. Then my foot finds a loose rock.

It isn't much of a slip, just a bobble, really, but it is enough. Sensing an opening, the plants make their move. One wraps itself around my waist from behind while another reaches over my shoulder from the front and attaches itself down the length of my spine. For a moment I want to laugh; it is an urge that passes quickly. The sun, after all, is terribly hot; sweat is beginning to flow in earnest; flies and other insects are hovering about my head; and these snaggle-toothed vegetables don't seem to want to let go.

The stalks are wrapped around me like coiled springs; the thorns are hooked through my shirt and into my skin. If I could just reach the tips of the stems I might be able to disentangle myself. But every movement anchors the plants more firmly, more painfully. Turn to the left and one tightens its hold; reach to the right and the other digs in deeper. I am unable either to progress or to retreat. I am trapped; I am scared. I might as well be lost in the desert: my chances of rescue are about as good. Insects begin to land around my eyes and on my lips. I feel in my chest the first fluttery stirrings of panic.

THE term *wilderness* is commonly associated with places far removed from our lives—with mountain vistas, forest depths, arctic silence. But wilderness is more than a matter of space and physical remoteness. It is spiritual isolation, mystery, danger, disorientation, and, most important, a sense of a power outside of and greater than oneself.

There is a wilderness adjacent to the world we daily inhabit, a hidden wilderness bound up in and for the most part obscured by the ordinary and the familiar. It can be seen through the grime on a city window; heard in the creaking of a wet limb on a raw, winter night; smelled in the wind that sweeps the suburban lawns at dusk. It is always there. It is part of the fabric of nature. But it is not often glimpsed until one is favored with a novel and unexpected view of everyday scenes.

"NOT until we are completely lost, or turned around . . . ," wrote Thoreau, "do we appreciate the vastness and strangeness of Nature." Likewise, to get lost—to find yourself in a familiar place that is suddenly, terrifyingly, unfamiliar—is one way to enter the hidden wilderness.

It is easy to do: Setting off along an oft-trod path the hiker decides, on impulse, to explore, to strike off boldly into the woods. Soon he learns that the spaces between the trails are bigger than he had imagined, big enough to contain false trails without number, trails that seem well-worn but which lead, not to human destinations, but to the secret haunts of deer and rabbits.

The paths grow fainter and fainter, and the hiker is stranded at last in a bramble thicket. He is calm at first, almost angry with himself. After all, he knows this place. He must be within a hundred yards of the road or the lake—if only he knew in which direction road or water lay. His heart beats faster as he turns from side to side, looking for landmarks, looking for help. The trees and the bushes close ranks around him, and the hiker, shivering, leans this way and that, starts, stops, returns, and starts again, suddenly desperate for the solidity of pavement and the sound of human voices. The leaves on the trees are rustling, but the hiker detects no breeze. Perhaps it occurs to him, as his last shred of composure dissolves, that people are not the only beings to have their moods and a limit to their tolerance; that they are not the only ones at times unprepared to welcome visitors.

THE limits of our experience in the world are much narrower than they may seem to us as we sit in our comfortable homes. In many ways we have cleared but a tiny path in nature's mystery; not an arm's length to either side there exists a realm in which other powers hold sway. It is a path limited in time as well as in space. Do not go too far into the woods. By the same token, do not stay out too late. To step off the path is to step into the hidden wilderness.

It is ten o'clock on a night in early spring. By starlight, the meadow path is a gray ribbon in the middle of a great blackness. Massed shrubs to either side seem like steep banks, and I walk as if in a ravine, under a pale strip of sky. It is too early in the year for katydids, crickets, and grasshoppers; the only sounds are the broken phrases of a restive mockingbird, and the lazy trilling of the chorus frogs in a nearby marsh. The misty trail, so familiar in the daytime, is now a path to nowhere, or to everywhere. Robbed of sight, I feel graceless and vulnerable. The flutter in the stomach, the faint fear, primitive and unquenchable, is a goading reminder that I am a diurnal creature; that any right I may have had to be here effectively expired with the setting of the sun.

THE HIDDEN WILDERNESS

The path soon leaves the meadow and enters the forest. The calling of the frogs fades into the distance, and the mockingbird falls silent, but as I move among the trees I become aware of another sound in the night—a stirring, a faint rustling among the leaf litter to either side of the trail.

The sound is not intermittent, but constant, and it follows me for more than a mile into the darkened woods. A hundred yards, a thousand yards, a mile. Each time I stop I hear the same gentle and insistent crackling from the ground at my feet. It is apparent that the surface of the earth is stirring throughout the length and breadth of the forest. It is as though last year's fallen leaves were coming to life in the starlight, wakening and stretching and flexing there beneath the trees, only to stop and lie still when the beam of my flashlight passes over them.

The sound is as uniform and pervasive as a gentle rain; as inexorable as growth itself. It is eerie. It is like being in a hall of a thousand doors, and all the doors are creaking but not one is seen to move.

Kneeling by the trail, I stare at the tiny circle of ground in the beam of my light. The rustling continues on every side. A minute passes. Two minutes. Three. Finally, with a tiny crackle, a leaf moves within the circle of light, slowly rises, then falls. I turn it over and find beneath it a pile of soft earth and the castings of an earthworm. It is a few moments before the full import of this discovery hits me. When it does, I stand and listen for a long time as the crackling, the rain-like patter proceeds in the night, on and on and on.

Worms. Good Lord, I think to myself, how many thousands? How many millions? Each night as we sleep they come from their holes to set the skin of the earth to churning. Across the darkened land the dead leaves lazily rise and fall as, beneath them, the world is turned inside out. Come daybreak the leaves are stilled; the worms descend into their burrows; our houses sit just a little deeper in the ground and we emerge, unsuspecting, onto the surface of a newly made planet.

THERE is wilderness in the most humble of places, yet you might spend all of your free hours out of doors and never stumble upon it. You may get hopelessly lost, walk all night long, and still discover nothing to shake your faith in the comprehensibility of the universe. Then, suddenly, it happens: the world slips, drops its guard, and you get a glimpse from the corner of your eye of something that, by all accounts, simply shouldn't be. The realization is forced upon you that you will not be privy to all secrets; that there is some play in the mechanism of nature, and not all is as fixed as we would like to imagine.

A day off from work, and I am out for a walk, as usual. It's my first visit to this particular park, and I feel a little out of place on the beaches and in the picnic areas, as if trespassing on someone else's turf. To make matters worse, it is a Saturday in the middle of summer, and people are swarming, as they will. The main trail is a broad thoroughfare made of gravel. Joggers and bicyclists pass in a seemingly endless file, disturbing the peace. When I see a narrow dirt path angling off to my right I take it without hesitation. Immediately I find myself alone, in a quieter, more congenial world.

 The path leads to a woodland swamp. Here are dark waters, dead trees, reflections of sky and clouds, anomalous roilings and ripplings, and bubbles bobbing to the surface—these, and a stillness, and repose, and the sense of creation ongoing. The light plays tricks here. It seems dim in the treetops, closer to its source, and brighter near the ground where it finds arching ferns, lush moss, and the leaves of skunk cabbage. Between green-rimmed pools are islands where the roots of straggling hemlocks break ground and grasp at their neighbors, trying to maintain their hold on dry land. Rotting logs are strewn all about, and white mushrooms gleam against the dark wood

 There is an indefinable aura about the tiny swamp, and all things coming into it are changed. A tiny rill is lively and clear until, cascading down a gentle slope, it finds the swamp and turns dark and still and brooding. Likewise the walker, coming from the light and the

THE OLD MARLBOROUGH ROAD

activity of the bike trail finds, on first touching the spongy ground, that he is walking more slowly than before, and looking over his shoulder, waiting for something to happen.

I am kneeling at the water's edge when from behind me comes a prolonged, staccato rustling, as of an animal scurrying through the fallen leaves. I turn, expecting to find a mouse or a shrew, and instead come face to face with a frog. Two inches long, olive green with a brilliant yellow throat, he sits there, next to the water, regarding me out of the corner of his eye with an expression that seems to say, "I've been here all along, an innocent, perfectly normal frog. No, you couldn't possibly have heard me run through the leaves because, as every educated person knows, frogs do not run." I hear the joggers laughing and talking on the nearby trail; I hear the rush of the passing bicycles; and yet there before me, in a gleaming amphibian eye, is the reflection of a wilderness independent of place.

WE normally think of wilderness as being vast, unpopulated, and silent. But it is something more. A lifeless city or an empty, million-acre parking lot would not stir in us the same feelings as, say, the Rockies or the North Woods. That's because wilderness is a matter not of space but of substance. It draws its power not from its size but from the stuff of which it is made—from animals and plants, from rocks and water and earth—and the smallest piece of that stuff, rightly seen, is as compelling and as enigmatic as the whole.

I remember standing one winter evening on a hill overlooking a lonely farm. At the base of the hill, on the bank of a frozen brook, stood the twisted, limbless trunk of a dead tree. I stared at it, astonished, knowing for the first time how it had grown there, seeing it in my mind as it pushed up out of the earth like the bony finger of a buried giant. That such massive beings could come into existence every day, all around us, without our ever giving them a second thought, I found remarkable. We do not understand what it is we are seeing when we look out our windows.

A leaf bursting from its bud; a dung-heap covered with flies; a weed pushing up through the sidewalk—a thinking person should quake before them. They embody the fecundity, the resilience, and the tenacity of all life. They imply power—power great and far-reaching, though cloaked in delicacy and an exalted slowness of pace. They hint at purpose, too, and answers beyond our ability to question. They point to the hidden wilderness that parallels our daily lives.

TRAPPED in the briers that summer afternoon, I am quite literally in the grip of a wilderness not a quarter-mile from human habitation. As I struggle, I picture the

THE OLD MARLBOROUGH ROAD

police (alerted, no doubt, by a hunter) disentangling my remains from the dead, brown stalks sometime in the fall. It is a sobering image, and it forces me to think. After some minutes' consideration of angles, slopes, and opposing forces, I bend myself into a pretzel-like shape and strain until (aha!) I can just reach the very tip of the stalk that grips my back. Then, closing my eyes and taking a deep breath, I give a convulsive outward yank. There follows an awful ripping and popping sound—and a wave of pain—and all at once I am spun like a top to freedom. A number of the smaller thorns remain embedded in my skin; they can be removed later. Spent, I stagger into the welcoming shade of a thicket.

There, at the edge of the clearing, I come upon two flies on a honeysuckle leaf. They are wasp mimics—thin-waisted, with conical abdomens banded in black and yellow. Their faces are gold, their eyes orange and oval. Their wings pump slowly; their abdomens rise and fall. They do not fly, even when I jostle their branch, for they are engrossed in feeding. I look more closely: they are eating bird droppings. Their swollen tongues are sunk into the wet, white mass like straws into a shared vanilla shake. On any other day I would probably stare in fascination, but something in my encounter with the briers has shaken me to the extent that I back hurriedly away and, breathing heavily, lower myself onto a log to regain my composure.

THE HIDDEN WILDERNESS

Arms crossed over my knees, head bent toward the ground, I catch a faint whiff of decay and notice at the same time a patch of black fur at my feet. It is a dead mole. A ring of tiny tentacles encircling its snout identifies it as a starnose, a species I have never seen before. Hesitantly, I poke the mole with the toe of my boot. And it moves. Its chest expands, and then contracts, and I fancy for a moment that it has come back to life. But then the entire carcass begins to writhe and palpitate. Ripples course back and forth under its skin. The mole belches a sudden cloud of ripeness and rot, and a greasy black beetle pops from its mouth and scuttles off into the leaf litter.

Choking, I lurch to my feet. Enough! Enough, already! For God's sake, let me out of here!

Perhaps it is just as well that the hidden wilderness is indeed hidden: constantly to be aware of the unspeakable potential throbbing within the least manifestations of nature might be more than we could bear. We live on the surface of our world in more ways than one, and just one glimpse of the realms beneath our tenuous perch is sufficient to make us uncertain of our footing forever after. Truly, we know not what manner of place we inhabit, only that it does not belong to us.

> . . . there is nothing to be commiserated,
> I do not commiserate, I congratulate you.
> —Walt Whitman,
> "To One Shortly To Die"

9
A Lesson in Dying

August 25: My thoughts turn to autumn, and the barest hint of the changing of the seasons is enough to stop me in my tracks. A single yellow leaf in the grass, a red creeper sprawled across a lichen-covered rock, a shower of leaves in the woods on a breezy day—they suggest the whole of what is to come: the Technicolor hillsides; the smell of burning leaves; hot cider on a cold night; Halloween, Thanksgiving and harvest festivals; corn shocks and fields of pumpkins; the gentle ripening and winding down of the world.

September 7: Yesterday afternoon, as I came home from work, the sun was blazing and the air was busy with the comings and goings of insects. Sometime

THE OLD MARLBOROUGH ROAD

during the night a great silence fell upon the land, and today, when I leave my house, I find myself in the midst of fall. The afternoon is cool and clearing after a morning rain, and the park is deserted. Distant hills look tired and worn, their uniform summer green fading toward yellow and brown. Up close, the woods are acquiring a more autumn-like look. There is a feeling of more space, more light between and under the trees. Sassafras is bursting into orange flame one leaf at a time; hickories are fading to yellow-brown; oaks, to red-brown. Virginia creepers are red spiral staircases encircling the trunks of trees.

I walk in a russet meadow studded with goldenrod and asters. No cicadas sing; no flies drone past; no grasshoppers clatter at my feet. The only reminder of summer's insect chorus is a sizzling trill so faint, so constant, and so pervasive that it might be the sound of life-energy itself, the quiet crackling of the potential hidden in the fading flowers, in the burrs and beggar-ticks clinging to my jeans.

There is a fall-like smell in the air, a hint of dust and ashes, a subtle, sweet decay. The gusty breeze has a newly sharpened edge, and the sunshine seems now more light than heat. Soon the last of the flowers will drop from their stems, and the songs of thrushes will be replaced by the rasping calls of crows and jays. Summer's soft drapery will fall away in tatters to reveal the sterner, more ascetic landscape that waits beneath.

September 25: We are in the gossamer days. Each morning the fog seethes, breaks apart, and drifts away, and the sun, burning through, reveals a wet world ablaze with color and bound in the embrace of spun silver. Over fences and bushes, from the branches of trees and the antennae of automobiles, the glistening threads stretch and hang and drape and shimmer tautly—orbs, arcs, and crystal lattices among the twigs; silver dollars strewn over lawns. The morning fields are bound in silver twine.

The webs I find are smaller versions of those encountered a few weeks back. They are the products of the newest generation, the spiderlings who, shortly after their birth, whether out of desperation or out of simple trust, loosed their silken lines to the wind, let go their grip on the earth and soared; who, bobbing at the ends of their filmy parachutes, drifted over forest, field, and town and bumped down here at last to make their ways in the world. Their tiny orbs cluster in the ends of branches; their tiny sheets encroach on one another in the grass.

October 1: Seated on the brink of Chickies Rock, I feel like the only stationary being in the world. In the course of a couple of minutes two dozen or more jays pass by far below, skimming the treetops in groups of five or six. They are uncharacteristically silent, intent upon reaching some unknown destination. Butterflies are moving, too: monarchs emerge one every minute or so from the woods behind me. Meeting the updrafts at the cliff's edge, they rise abruptly, teetering, then steady themselves for the long shallow glide over the railroad tracks, over the river, over the far ridge and toward the Yucatán.

Even things without wings are moving. Woolly bears and other caterpillars are out in numbers, hurrying across roads and trails, looking for places to hibernate. The air over the river is filled with spiderlings on their silken draglines. The shimmering threads drift slowly toward the rock and then, catching the updrafts, shoot up the cliff face and sling their tiny passengers over my head toward the ridge-top. I feel like taking a walk. October is the month for vagabonds.

October 2: There is something about the fall that infects animals with a certain strangeness. Lately odd whistles and creaks turn out to be coming from blue jays and flickers. And chipmunks, for reasons unknown, have taken to clucking like turkeys from the mouths of their burrows.

As for the squirrels, they are driven these days by a sense of urgency bordering on hysteria. A few weeks ago they would simply stand and look as I drove by. Now they dash off, leap ditches, spring onto stumps,

A LESSON IN DYING

jump back down, and tear off into the woods, scattering the leaves behind them. Their skittishness is understandable, I suppose. It is, after all, a busy season, with serious provisioning to be attended to. I watch through binoculars as a squirrel sprints along the top rail of a fence. Coming abreast of my position, he suddenly stops and sits up. Breathing heavily, he claps a tiny paw to his chest. After a moment's rest he drops back down to all fours and takes off running again.

For all their seeming anxiety, the squirrels need not fear starvation this season. I have never seen a more massive acorn crop. Nuts lie in carpets under the trees and form deep windrows in the hollows between the roots. To pass near oaks is to walk on marbles. Twice my feet fly out from under me. Once I am thrown against a tree, and once over an embankment into the mud.

Fall is feasting time, stuff-your-face time. Rotund woodchucks are busy in the fields and at the roadsides, laying on winter fat. When startled they waddle toward their burrows, bellies scraping the ground. Fruit abounds, and one is wise to keep an eye on low-flying birds. Purple splotches dot the cars, the sidewalks, the grass.

I am foraging, too, along with the rest. No nuts and berries for me, though. I seek to harvest an earthier crop. It's meadow mushrooms and puffballs I am after,

THE OLD MARLBOROUGH ROAD

and sulphur shelf and tree ears and boletes—fungus, the lover of dead wood and loamy earth. It is a food with ritual value, as symbolic of the season for me as pumpkin pie or apple cider might be for others. To eat wild mushrooms is to eat damp leaves and old stumps once removed; it is to taste the autumn, to take a bite out of the rich, ripe, rotting fall earth itself.

A few weeks ago, on a damp, drizzly day, I followed what seemed to be a flame in the gray woods and came upon a clump of mushrooms growing from the base of a stump. It was the jack-o'-lantern, a poisonous species named not for its brilliant colors but for its purported habit of glowing in the dark. I took some home to see for myself. Carrying it into a darkened room, I placed it on the table before me. Nothing seemed to happen at first. And then, as my eyes grew accustomed to the darkness, the fungus began to glow. Its gills shone with the faintest and coldest of lights, a light too dim, too unearthly, to have a describable color, a true autumn light, a light that mocked the idea of light as we know it.

October 10: In the woods the silence grows day by day. Every now and then a chipmunk will squeal and rustle the leaves, or a late-staying bird will take wing and turn southward, but the earth has borne her young—her seeds, fruit, and spores; her spiderlings and caterpillars. She has scattered them abroad or stored them safely away, and now she rests, and waits, quiet and alert—poised, it seems, for some momentous transformation. I wait with her. I watch and I listen. In the fall I am effortlessly attentive. I drift across the land like smoke, filled with uncertainty yet trembling with an elation whose source I cannot name.

October 14: I stand outside a farmhouse. The last of the light is a swath of purple above the mountains to the west. The night is cold, and a stiff breeze from out of those hills is rolling clouds across the sky like balls of black cotton. Gradually, I become aware of movement in the darkness; of running footsteps coming from the mountains, from where nobody lives, approaching over the grass and the gravel driveway to the west. Whoever it is is moving fast. The footfalls thump at an incredible pace, getting closer and closer, and yet I can see no one in the dim light.

When I was young, and Halloween was coming, I thrilled to stories about ghosts and goblins and demon spirits. I rode with witches on their brooms, and accompanied Ichabod Crane on his mad flight through the midnight woods toward the old bridge. On Halloween itself, walking the small-town streets in my costume, I sniffed eagerly at the autumn air, shuffled through the crackling leaves gathered at the curb, and felt the proximity of a dark presence that was in equal measure dreadful and compelling.

THE OLD MARLBOROUGH ROAD

Now I realize that this presence was not something conjured up by the stories and the costumes, but rather something inherent in the season itself. Now, as a phantom runner approaches from empty mountains in the dusk, I understand that what I really took away with me on those childhood Halloweens was not apples and lollipops but some inkling of all that is implied in the word *mortality*.

The wind thrashes at the trees and thrums the boards of the old house; the clouds tumble past overhead. The steps, still unaccompanied by any visible form, grow louder, heavier, closer, and then, as I brace myself to meet the disembodied guest, a sprig of dead leaves materializes at my feet and tumbles past, clattering, in the wind. My heart is pounding as the sound of running feet fades off into the fields at my back.

The autumn is a traveling circus in an out-of-the-way town at dusk, behind whose laughter and colored lights there looms the shadow of something sinister. It is a season of latent meaning, when our nightmares take on substance, detach themselves from the shadows at twilight, and confront us when we least expect it.

October 16: At the park on a gloomy and threatening afternoon. The skies and the lake are the color of pewter. A strong wind whips up waves on the lake's surface; it sweeps leaves from the treetops and carries them high over the dark water against the overcast.

A LESSON IN DYING

They swirl and tumble and seem to sail in slow motion, like a cloud of drab moths. A flock of grackles wheels in over the lake, then banks and funnels downward, rushing through the treetops and away like a black wind. The oaks in the woods are hanging onto their dead leaves; they make a dry and throaty rattle. The fancy strikes me that it is the death rattle of the earth; that the world is dying, not in some metaphorical sense but here and now before me; that I stand by its death bed; that this is the moment, regardless of what the calendar says, when autumn reaches its culmination.

Suddenly the rattle ceases; the end has come. As the sun sets, a stray beam finds a rent in the clouds and the land about me catches fire. It is a light that tugs at the heartstrings. It filters through the trees like powdered gold, and flows across the grass and roads like melted butter. The black water turns blue; the gray waves are pink. The falling leaves turn from moths into butterflies, and a maple on the far shore seems to burst into flame over the shadowed lawn. Everything the light touches—the least twig, the tiniest pebble, the most ragged dead leaf— glows in splendor. The landscape shines for a moment like the near edge of paradise.

There are those who have glimpsed in October the promise of another April; who have found in autumn's sowing and scattering and provisioning a guarantee of renewal and rebirth. But as the sun sets, and the light fades, and the silence settles upon this lonely lakeshore, I am struck by a far more forceful vision. The world is dying, yet it has never been so lovely. The world is dying, and all's right with the world.

If autumn promises anything, it is not that there will come a spring but that there will come a winter.

If autumn speaks at all, it is not of resurrection but of rest.

If autumn is poignant, it is not because the world is passing away, but because we are being left behind.

October 23: Home. I step outside into the night. The forecast calls for frost, and at ten o'clock I am shivering in my light jacket. One or two crickets are still active, but they sound very weak, faint, and far away. It is as if they were slowly drawing off, retreating back into that dark, hidden center from which all things arise, leaving me alone, in the silence and the bitter cold.

> To be alive—is Power—
> Existence—in itself—
> Without a further function—
> Omnipotence—Enough—
> —Emily Dickinson, #67

10

October Firefly

IN 1990 people in the United States and elsewhere celebrated the twentieth anniversary of Earth Day. Suddenly, it seemed, environmentalism had become fashionable. Television specials featured the icons of pop culture cavorting with whales and porpoises, or reporting on environmental problems, via satellite, from around the world. Network news programs devoted lengthy (four- to five-minute) segments to exploring the state and eventual fate of the planet. The papers ran special sections with titles like "Ecowatch" and "Earthview '90," and covered a different environmental issue each day. For a week or two, people massed and

THE OLD MARLBOROUGH ROAD

marched and chanted and signed petitions. "Save the Earth!" they cried, "Save the Earth!" To which I respond, respectfully and politely, "Humbug." For as much as I love the forests, the fields, the lakes, the rivers, the swamps, the deserts, the mountains, and the seashore; as much as I cherish deer, bears, trout, bluebirds, rattlesnakes, spiders, salamanders, jellyfish, toadstools, dandelions, oaks, maples, orchids, dragonflies, and earthworms; as much as I applaud the sincere efforts of the individuals and organizations striving to preserve wild nature, I wince when I hear people talk of "saving the earth."

In the first place, the earth has no need of salvation. In fact, the world, I suspect, takes us far less seriously than we take ourselves. We shudder, and rightly so, when we consider the potentially devastating consequences of deforestation, ozone depletion, hydrocarbon emissions, the dumping of hazardous wastes, and all the other environmental idiocies in which we are prone to engage. But it does not necessarily follow that the planet shares our feelings. In *The Fate of the Earth*, Jonathan Schell describes something close to a worst-case scenario: a prolonged, intercontinental thermonuclear exchange. In the wake of such an occurrence, he writes, the United States could be transformed into "a republic of insects and grass." This prospect understandably terrifies us; yet there is no reason to believe that the earth would be much distressed by such a turn of events—providing, of course, that the grasses were lush and the insects healthy and abundant.

For the earth is bountiful but not sentimental. Her concern is not for human life, or the life of any given species, but for life in general, life itself. And life, I am convinced, is a flame that will not be extinguished by the likes of us. It was first kindled upon a planet more barren and inhospitable than we can imagine, and it has burned on for billions of years in the face of climatic change, worldwide glaciation, volcanic eruptions, continental collisions, even meteoric impacts so tremendous as to annihilate up to 90 percent of all species in existence at the time.

If we continue on our present course we may leave the earth raped and gouged and charred and poisoned and possibly bereft of all but the simplest organisms—but in the process we will seal our own doom. And when we have gone the earth's smoldering life-fire will crackle into flame once again. Living things will multiply, evolve, and spread. Unfamiliar animals will scurry among the novel vines and creepers twining themselves about the ruins of our once proud civilization. And it will be no less wondrous than before, only different.

In the second place, I am convinced that the vast majority of those who cry "Save the Earth" are unpre-

THE OLD MARLBOROUGH ROAD

pared to carry this sentiment to its logical conclusion. Take for instance the smallpox virus. It is now, by all accounts, extinct in the wild, and humanity is responsible for its demise. Are we now to hear calls for its reintroduction? Similarly, are we soon to see marches and demonstrations protesting attempts to eradicate mosquitoes, black flies, locusts, and other disease-carrying or crop-destroying insects? I would venture to say it is unlikely. Yet these creatures are as much a part of the earth as whales, or rain forests, or people.

My argument, of course, is merely one of semantics, yet I think it is valid. Today, people in unprecedented numbers are acting to address urgent environmental issues. Many of us are wasting less and conserving more, demanding clean air and safe water, fighting for the preservation of open space, and speaking out on behalf of animals, plants, and entire ecosystems unable to speak for themselves. These actions are commendable, and necessary, but to categorize them as "saving the earth" is to distract attention from what I think is the actual, if unstated, goal. What most of us are after, I respectfully submit, is not to save the earth per se. What we are after—and it is the most natural desire I can think of—is to save the earth we know, the earth we love, the earth we need. What we are after, in the end, is to save ourselves.

The question is, can even this much be done? In the

long run, probably not. In the long run, extinction, in one way or another, is probably inevitable. Why, after all, should our species alone be exempt? As for the more immediate future, the one or two generations next to come, the world that will be our children's and grandchildren's—the jury is still out.

IN the study of ecology there is a concept known as the edge effect. It holds, simply, that you will find a greater variety of organisms at the boundary between two habitats than in either habitat by itself. In my line of work one discovers early on that there is an edge effect in time as well as in space; that it is neither the day nor the night, but rather the twilight, that is most congenial to the denizens of the mind and the imagination—to apparitions, dreams, visions, and possibilities.

It seems today that the entire world is bathed in the cold blue of some greater twilight, and possibilities, both encouraging and dire, are unfolding at every turn. Today, modern communications and transportation allow the most far-flung cultures to meet and understand one another; personal computers permit each of us to tap into civilization's collective store of knowledge without leaving our chairs; advances in medicine and medical technology make it possible for infertile couples to have children, for premature babies to survive, and for the victims of accidents and disease to recover and resume their lives. On the other hand, many people are homeless and hungry, even in our most prosperous cities; our young are falling prey to drugs, alcohol, depression, and suicide; lakes and rivers are turning acidic; seas shimmer with oil; rain forests are going up in smoke; and the elephants, the whales, and the great apes are in danger of annihilation.

In many ways human life has never been better; in many others it has never seemed to be in such acute peril. It is difficult to predict to which side the balance will eventually tip.

Most of the people we know are decent people. They want the best for themselves, for others, and for the planet. Their intentions are good. Yet things do not always turn out as we would like. The threads that link us to our world and to one another have become so intricately interwoven that a single evil individual can throw civilization into turmoil, and people en masse set in motion forces that take on lives of their own until no individual or even nation of individuals can predict their consequences, let alone hope to control them.

It seems at times as if there were a warped and spiteful presence hovering about us, taking our every action and turning it sideways, thwarting our best intentions. We pluck what we think is a tiny bit of lint from the world's lapel, and the whole fabric of things unravels. We find ourselves poisoning the land with pesticides to increase the world's supply of food; taking meat from hungry, third-world families to protect endangered mammals and birds; capturing condors and locking them in cages so that they may someday fly free once again.

Amoeba-like, civilization crawls. We hope that, collectively, we are moving forward; that things will somehow work themselves out; that our wisdom and our foresight will prove stronger than our greed and our capacity for thoughtless tinkering; that our children, and their children, will be heirs to a planet which, if it is not better than the one we know today, is at least no worse. But our chances would appear marginal at best. We will not know where we are going until we get there.

All about us, the half-light lingers. It is difficult to say if it is the light of dawn or of dusk.

The crowning irony is that even if we succeed—even if we manage to maintain the earth as a fit home for both humans and other creatures—our work will not be finished. Rather, there will remain to us one final task—a task that will fall not to society, not to the species, but to each man and woman alone. For no matter how hard we may strive to pull civilization back from the shadows, each of us is destined at last to confront a personal twilight.

I LOOK again at the picture on the desk, the picture of my father and me on that rutted mountain road. I was very young, but I remember those walks, especially the ones taken in the fall. I remember the slant of sunlight through the trees; the deer, briefly glimpsed, bounding away up the hillside; the leaves, their colors, and their crunch, and the way they smelled when you lay sprawled among them. I remember my impatience at being made to sit still in the hope that the deer would come back, when what I wanted to do was shuffle off through those orange and brown and red drifts into the heart of the woods, to explore.

I remember one of those walks particularly well. I had decided that I was going to be a paleontologist when I grew up, and I was collecting rocks that showed any sort of little marks on them, any evidence of containing fossilized creatures. Collecting was going especially well that day, and as my bag grew heavier, my father began to get impatient, for he knew all my specimens were headed for the shoe box under the bed at home—a box already half-filled with unremarkable gray and brown and white shards of stone, collected mostly from the backyard.

The day's hike was just about over when I discovered the most incredible specimen of all. Just off the road, next to a big gray stump, there lay a flat stone perhaps eight inches across. I picked it up, brushed a coating of dried mud from its surface and found myself staring at a footprint—a large and (I was sure) reptilian footprint, the footprint of a dinosaur.

Trembling, I opened my bag and was putting the stone inside when Dad laid his hand on my wrist. No, he said, the rock was too big, and I had done enough collecting for one day. I tried to tell him that this was the find of a lifetime, but I could barely speak. I tried to explain that no matter what else I left behind this one rock had to come home with me, but it was no use. I don't remember if he even looked at the stone. If he did, he couldn't see the impression that to me was so clear. "Put it down," he said. "We're going home." I was crying when I laid the stone next to the stump, and as we walked off I looked back again and again. I tried to impress on my memory exactly where I left it, but I knew that even if we came back I would never find it again.

I could not have explained it to my father—I can't really explain it now—but that seemingly insignificant bit of rock was something I was meant to have. Even today, when I know that the chances of my having found a dinosaur footprint in the shales of Pennsylvania are essentially nil, I occasionally find myself picking up and running my hands across stones of a certain size and shape. That stone still lies there on the fallen leaves of memory, bathed in a brilliant light, beckoning. It is a

THE OLD MARLBOROUGH ROAD

part of my world that does not exist in the worlds of others.

The truth is that however much our lives may seem tied to the lives of those around us, for each of us there is a landscape of the soul that must be traversed alone. Though born by chance into families, races, nations, we inhabit different worlds—worlds that overlap only at their outermost edges, and whose centers are dark and unexplored. That end of the world we so dread is not some abstraction, something located in the indefinite future, to be faced by our children, and our children's children. The end of the world is as close as the next malfunctioning traffic light, the next false step at the top of the stairs. We face the end of the world every moment that we live. Our immediate goal, as a group, may be the preservation of our species, but our ultimate task as individuals—if we will admit it—is coming to terms with the bittersweet evanescence of our own existence.

MY daughter is in the next room, looking out the window and carrying on a loud, unintelligible conversation with something outside. It might be a passerby, or a bird, or even a cloud for all I know. She is just three years old and already our paths are diverging; already she has set off along that road which is hers alone to travel.

No doubt many of my worries on her account are unfounded. It is true that she may grow up in what I would consider a bleak and urbanized world. But people have a remarkable capacity for adapting to and accepting the world into which they are born; for looking back on the times and places of youth with fondness, whatever those times and places may have been. I'm sure my father would have preferred that I grow up in the kind of world he knew as a boy, yet I am content in the world I know. I might look back on past eras with a feeling of wistfulness, wanting to experience them, but I am not unhappy now; I might regret never having seen a passenger pigeon, but I am not losing sleep over the fact. In the same way, though I myself would not like to live in the sort of sterile, technological future envisioned by the prognosticators, to those born into that world it will be home.

Still, life is difficult even in the best of worlds. My little girl, on her journey, is apt to pass through dark and forbidding reaches, and there is only so much I can do to ensure that she will eventually find her way.

I would like to give her a comfortable home and a good education, of course, but such will serve only to start her out in the right direction. I would also like to leave her a society that has found within itself the willingness to share the planet with tropical forests,

THE OLD MARLBOROUGH ROAD

spotted owls, and the rest of wild nature. But such willingness, as we have seen, may or may not be forthcoming.

I suppose if I could leave my daughter one bequest it would be an antidote to despair, a touchstone of strength and hope, a reason for persevering in the face of *whatever* may eventually come to pass: it would be, I think, a vision I was granted many years ago.

It was an evening in late October, brisk but not yet freezing, and I was walking home just after sunset. Night was coming on, the darkness rising up at my back and hanging overhead like a wave, but the western sky in front of me still held the deep, bottomless blue that seems the exclusive property of fall and winter evenings. Bare trees in silhouette on the horizon looked like delicate gorgonian corals anchored to a bit of rock deep in a storybook sea. There was no wind, and all was quiet save for the occasional car and the crunch of fallen leaves as I passed.

As usual, I took the short cut through the cemetery. A light mist had begun to rise and curl around the knees of the pallid stone figures on their pedestals. I had stopped to take in the peace and the stillness, to brood a while in the last of the light when, incredibly, just in front of me, there appeared a firefly, a little green meteor that rose through the mist and then winked out. He flashed again a little farther off, and then again, receding into the night—searching for a response from the grass below. It was a response that I knew would never come. Adult females of his species had been dead for weeks, and only the young remained, wormlike larvae nosing about among the grass blades or steering by their own cold light through subterranean tunnels, preparing for the winter.

Something somewhere had gone awry. The firefly had slipped through a tear in the fabric of nature, and now was utterly alone in the world, out of sync with the rest of his kind, somehow dislocated in time. He would never find his way back, never make the contact he sought. One night soon the frost would come, a hard frost, and that would finish him. He flashed once again—his light all the more brilliant and beautiful between the mist and the darkening sky—then disappeared for good into the dying of the day and the dying of the year.

The firefly could not have known it, I reflected, but the flight, that to him must have seemed so cold and lonely and in vain, was to me an act brave and brilliant and achingly, inexpressibly lovely. It struck me all at once that I had been granted a great gift, for I had been shown my life, a human life, from the perspective of a god.

It is this vision that I would leave with my little girl. In this world there is ample opportunity for disappointment, disillusionment, anxiety, confusion, and pain—but there is never cause to lose heart. For the strength to persevere is as close as the grass under our shoes, the rain on our cheeks, and the wind in our hair. It springs from the recognition and remembrance of what it means to have been granted the opportunity to live.

We are all October fireflies, drifting through the dusk toward unknown, private ends. But if our flights are brief then they are also brilliant. We are each of us citizens of equal standing in a wide, wide universe, and all our laughter and love, all our fear and sorrow and loneliness and longing pale alongside the glorious improbability of existence itself. Our least actions unfold in the light of a billion galaxies. If we forget this, no Utopia we can fashion will bring us joy; if we remember it, then life, whatever its circumstances, cannot help but be a journey into wonder.

Eventually, that last long night will be upon us. Eventually, there will no longer be "more day to dawn." Well, what of it? For dawn is a gift, but darkness, in one form or another, is guaranteed. And the darkness may hold treasures greater than we can imagine. It is only at night, after all, that we can see the stars. So when, at last, the air grows chill and the shadows long, let us not cower and curse our fate. Rather, let us rise and feel the night wind under our wings. Let us kindle our little lights, and let us shelter them. Let us savor the sunset while it lasts.